Grumman A-6
Intruder

2195

OSPREY AIR COMBAT

Grumman A-6
Intruder

Robert F Dorr

Published in 1987 by Osprey Publishing Limited
27A Floral Street, London WC2E 9DP
Member company of the George Philip Group

Sole distributors for the USA

Publishers & Wholesalers Inc

Osceola, Wisconsin 54020, USA

British Library Cataloguing in Publication Data

Dorr, Robert F.
    Grumman A-6 Intruder.—(Osprey air combat).
    1. Intruder (Bomber)
    I. Title
    623.74′63   UG1242.B6
    ISBN 0-85045-816-1

Editor Dennis Baldry
Designed by Gwyn Lewis

Filmset and printed in Great Britain by
BAS Printers Limited, Over Wallop, Hampshire

FRONT COVER
*Still painted in full squadron colours and fully laden with
external fuel tanks, a KA-6D Intruder from VA-95 'Green
Lizards' lines up for a landing over the creamy wake of
Enterprise in June 1986. Capable of carrying 2,500 Imp gal
of JP-5, the tanker version of the A-6 is the only truly role
dedicated in-flight refuelling aircraft in the Air Wing. When
an Intruder squadron is embarked for a cruise, four or five
KA-6D tankers will usually be included in the 15–16 aircraft
total*
(Tony Holmes)

TITLE PAGES
*Bombs away. An A-6 Intruder, buno 152946, from VA-75
'Sunday Punchers' hits back with a US Air Force F-4D
Phantom II from the 8th TFW (lead) and a pair of Marine
Corps F-4Js from VMFA-232 during a Linebacker I mission
over North Vietnam in July 1972*
(via Larry Davis)

Contents

For Lew Abrams

Introduction

Braking the car to enter town on Route 235, the busy two-lane highway is dotted with signposts of the American fast-food affliction (*Pizza Hut, Mister Donuts*), the signs providing the only colour in the wet grey Maryland afternoon. There are woods inside the base's perimeter fence over to the left, and above slickened treetops in a dull grey sky, a dull grey Intruder banks in a lazy circle. A left turn at the only crossroads in this backwater town and, after security checks, the car is waved through a well-guarded military gateway. There are not merely trees but a golf course (perhaps there is a golf course at every major installation of the United States Navy). The disorder and ticky-tack of the town outside—the kind of small, tight-knit town of straight and sturdy people who go off to fight our wars—gives way to the clipped grass and pristine, painted buildings inside the base. It is a truism that Americans who excel in battle come not from the Ivy League eastern establishment but from rural hamlets like this one, but this town is different because its military base, its life's blood, is not merely another installation. It is the Naval Air Test Center at Patuxent River, Maryland, where the best and brightest have wrung out our warplanes and sometimes given their lives for several generations. A sign warns, PHOTOGRAPHY RESTRICTED TO RECREATION AREAS ONLY. The Intruder is still circling.

I had seen the Grumman A-6 Intruder before, of course. In 1968, newlywed, with North Korean commandos firing rounds over my house in Seoul, *Ranger* and *Enterprise* came to our support with braces of Intruders on board and the bad guys backed down. In 1985, with a Prairie Fire smouldering further south, I went aboard *America* off Portsmouth, dined with the skipper, and sat in an Intruder cockpit. At dozens of air shows and open days, even at Patuxent where the sun *does* shine as much as elsewhere, I had seen Intruders and Prowlers and given them exceedingly little notice. The Intruder does not attract attention. Its crews do not seek glory. It is an ugly, pragmatic machine flown by modest, pragmatic men who talk little about their trade but go out, when we need them to, to fly and fight.

A brief drive through the base leads down to Strike Test, where American naval aviators, risen to become among the world's finest test pilots, put Hornets and Tomcats through their paces. And Intruders—even the unsung Intruders which are now approaching thirty years of age. At Strike, they are working on the Intruder which in rejuvenated form will serve the Fleet for the remainder of the century.

The F-14D and A-6F Program Office rests in a low-slung building across from the flight line. The office is headed by Captain Ken (Doc) Koskella. A peculiar background, Koskella's, to be bringing the next-generation A-6F Intruder into the world: he is a flight surgeon as well as an attack pilot. 'The Doc' is in charge of several dozen men who are pushing today's Intruders, not to the edge of the envelope, but to the edge of newfound capabilities, and who will be test-flying new A-6F Intruders by the time this is read. 'The Doc' is calm, businesslike, perhaps mildly bored while Lieutenant Jim Gigliotti, bombardier-navigator, discusses the Intruder. The Navy's training units come up in the conversation and one of Gigliotti's fellow fliers is annoyed at a European publication which insists that references to Replacement Air Groups (RAGs) are in error now that the Navy calls them Fleet Replenishment Squadrons (FRSs). 'We don't say that. We call them RAGs.' Gigliotti describes transitioning from the TC-4C Academe trainer to the A-6E TRAM Intruder. 'You go out to the Fleet for two-and-a-half, maybe three years, you get crewed up with a more experienced individual, and maybe you'll do two cruises in that period with up to a hundred carrier landings . . .' Not talkers, not comfortable discussing their airplane and their craft, Koskella and Gigliotti would rather be flying, even in today's murk. Murk, after all means nothing to the Intruder, which can fly and fight when even ducks refuse go aloft.

A history of a combat aircraft is supposed to include mention of its faults, as much as its merit. The author can find no fault with the Intruder—not, at least, with the exception of the difficulties with internal systems which were endemic when the airplane was first rushed into combat. When beseeched to find one thing, just one thing, that is *wrong* with the Intruder so that the author can mention a weak point along with the strong, Gigliotti is speechless. Finally, Doc Koskella allows as how the nosewheel steering is too sensitive. 'You have to hold the nosewheel steering button down when taxying and it's over-responsive.'

Lieutenant Dave (Cleve) Brown, ex-VA-65, ex-USS *Eisenhower*, is prevailed upon to let the author actually see and crawl over an Intruder. Chosen by happenstance is the latest one in the Navy, A-6E-90-GR TRAM Intruder 162182, modex SD, side number 505. The airframe after this one on the factory line is the first A-6F. Dave Brown—an individual's tactical callsign, Cleve in this instance, does not arise in everyday chatter—talks flying with his hands. Overhead, that circling grey Intruder makes another banking pass over Patuxent in the grey day and he looks up. 'I always liked airplanes.' Until Dave pointed it out, I never knew that you can actually climb up *inside* the avionics bay in the rear of the Intruder, the 'birdcage'. Until Dave pointed it out, I did not know that the yellow handle will get rid of the canopy, although you can eject *through* the glass if you have to. Until Dave explained, I did not know that the radar warning receiver gives you a warbling in your earphones when they're locking onto you with a *Fan Song* radar. Dave even has a comment about the red and white handle, which is secured with a lead seal and criss-crossing wire. 'That's nuclear.'

The Grumman A-6 Intruder remains something of an enigma to this author. The Intruder, and its Prowler stablemate, can make no claim to beauty. It has no constituency. It is an ugly, practical, supremely pragmatic aircraft design which seems to have more longevity than any of its contemporaries. It is not merely in production, it is being improved—hence the low-slung building across from Strike Test where Doc, Jim and Dave do their thing. And it is an enigma because, somehow, the quintessential Intruder is both hard to find and difficult to define. The narrative which follows is but the beginning of a story which will range into the next century.

Others have already told portions of the story and it would be remiss not to say, here, that there exist fine publications about the Intruder and Prowler from the typewriters of Lou Drendel, Bert Kinzey, Roger Chesneau, James Wogstad and Phillip Friddell. The story is big enough for all of us and it is a continuing saga, as one Libyan tent-dweller learned in recent times.

This modest effort to continue the Intruder saga will inevitably contain some glitches, as all attempts at history must. Any errors are the sole responsibility of the author. But this volume would have been impossible without the generous assistance of many.

On 24 November 1967, Lieutenant Colonel Lew Abrams, skipper of Marine All Weather Attack Squadron 242, set forth to go downtown in A-6A Intruder 152612 and, with his bombardier-navigator, failed to return. In remembrance of all those who took Intruders to Hanoi, to Beirut, to Benghazi, this volume is dedicated to Lew Abrams, all honour to his name.

I am again in debt to editor Dennis Baldry who by now is indisputably the best in the business. In an era when too few aerospace firms show a concern for history, Lois Lovisolo of the extraordinary Grumman History Center provided invaluable help, and much-needed counsel, supported by colleagues H J (Schoney) Schonenberg and William C Barto. Even Young Soon Dorr helped out a little.

Assistance in the preparation of this work was received from the Department of State, the United States armed forces, and numerous American military units. The Grumman Corporation provided help at every juncture. Encouragement also came from Lt Col John D Cummings, soon to become the newest bombardier-navigator in the United States Marine Corps. In this one, John, the other guy sits to your left.

I especially want to thank David A Anderton, Hal Andrews, Paul Auerswald, LCDR Richard Burgess, Dr Alan Curry, Jack Elliott, Dick Feyk, William Green, Bill Gunston, Joseph G Handelman DDS, Capt Paul Higgins, Martin Judge, M J Kasiuba, Joseph Le Strange, Don Linn, Peter Mersky, Cdr Bob McHale, David W Menard, Lt Col Robert J Mills Jr, David Ostrowski, Dave Parsons, LCDR Tom Patterson, Chris Pocock, Norman Polmar, Brian Salisbury, Norman Taylor and Nick Williams.

The views expressed in this book are mine and do not necessarily reflect those of the Department of State or of the United States Air Force.

Robert F Dorr
Oakton, Virginia, June 1987

Chapter 1
Intruder Genesis

'Any fighter guy could tell you, the pointy end is supposed to be at the *front*.' With this put-down, an Air Force officer dismisses the Grumman A-6 Intruder. Or, at least, tries to. But the Air Force could do worse than to possess a few of the *un*pointed attack airplanes which emanate from the creative mind and engineer's pencil of designer Lawrence M (Larry) Mead and from the incomparably experienced Long Island manufacturing firm founded by Leroy R Grumman. The Navy and Marine Corps, it will be recalled, have employed the Intruder to good effect for three decades and have flown it in combat in Vietnam, Lebanon and Libya. Admit it or not, there have been a few times when the Air Force could have used such an aircraft. On 13 January 1987, I flew up to the Grumman plant at Bethpage to find out why.

Origin of an Intruder

The A-6 Intruder came into the world to fill a need, evidenced during the Korean War, for an attack aircraft able to strike targets at night in bad weather. That's what they all say, those Intruder stories which reveal but a part of the human background to the flying hardware of today. What they write is true indeed. In fact, in Korea some of us would have reached into our own wallets if that would have bought us a night or bad-weather attack capability. But while browsing through the personal papers of Larry Mead at Grumman in Bethpage, I could not help but observe that there is so much more to the Intruder story.

Mead was away on a trip. I had missed by days an opportunity to interview the Intruder's designer. But his papers from three decades ago told a poignant and intricate tale.

In mid-1956, American aircraft manufacturers were invited to respond to a Request for Proposals (RFP) known as Type Specification 149 or TS 149. The US Navy's written statement of its needs was not especially detailed but did establish certain requirements. The attack aircraft built in response to the RFP was to have a two-man crew, all-weather capability, a radius of 300 nautical miles (500 km) for close-support and 1,000 nautical miles (1500 km) for long-range interdiction, with speed of 500 knots (800 km/h). The new aircraft would of course be built for carrier deck operation and apparently it was necessary to be mindful of both the hydraulic catapults aboard *Essex*-class carriers and the steam catapults to be found aboard a new class of supercarrier. TS 149 also specified STOL (short takeoff and landing) capability, apparently in the expectation that the new aircraft would be operated by the Marine Corps close to its amphibious beachheads.

This last requirement may have caused more concern than the others to Larry Mead, the head of the Grumman design team responding to the RFP. Now, a generation later, I was given access to the personal papers which revealed how Mead and others in the team set forth to create an Intruder: I could see, at a glance, that various approaches had been taken to fulfilling the sparse requirements of the RFP and to anticipating the obvious. Important design elements were the massive search and track radars needed for the all-weather role, the two-man crew, and a wing of low wetted area and moderate sweepback. Now, although looking at Mead's personal memorabilia with the blessing of its appointed custodian, Grumman corporate historian Lois Lovisolo, I could not escape the feeling that I, too, was an 'intruder' of a kind: it was an honour to peruse raw work by one of the world's leading aircraft design people.

It was all there in a cardboard box—the early designs for an M-winged attack airplane which was never built but which, in the design shop, evolved into the Intruder; the later designs for the Intruder itself in versions which reveal a process of evolution but were never actually built. Mead's paperwork even includes the original Grumman letter accepting the Navy's offer to bid on an attack design. The letter

In natural finish, A2F-1 prototype (147864) is taken for its first flight by pilot Robert Smyth on 19 April 1960. Smyth never did raise the landing gear. During later tests, however, 147864 acquired standard US Navy gray and white paint and became the Intruder programme's spin-test airframe (Grumman)

RIGHT
The full-scale mockup of the A2F-1 Intruder is seen at the Grumman plant on Long Island in 1959 after Navy contract was awarded. The mockup was displayed to Navy inspectors with various fuel and ordnance loads, including (not shown here) the Corvus missile which was later cancelled (Grumman)

was typed 'in final' but someone, apparently Leroy Grumman himself, pencilled in minor changes and the letter had to be retyped. The 'final' copy with the pencilled notes, thirty years old, is among Mead's treasures. Of this, more below.

It must seem incredible now, the notion that today's Intruder might have come out resembling an Ilyushin Il-28 but with an M-shaped wing, swept forward inboard of the wing-mounted engines, swept back outboard. The evidence is there. Mead and his design team were sketching out M-winged attack aircraft immediately after the issuance of TS 149 and well before the manufacturer's actual response to the RFP. Although today's A-6 is known in Grumman company parlance as Design 128, Mead's pencil created an entire series of M-winged attack airplanes which employ 'concept numbers' in the 128 range, leading to some duplication. A glimpse at preliminary drawings indicates that Mead was examining a number of M-wing attack craft. A sampling:

Mark: September 1956
Concept 128F—its exact relationship to the story

G-59100

High over Long Island, the much-tested number-one A2F-1 Intruder (147864) performs stall tests. The starboard wing and tailplane of the Intruder have been rigged with tufts to enable the camera to record airflow over these surfaces (Grumman)

LEFT
A number of variations on this basic M-winged configuration were sketched out by Larry Mead's design shop before the M-wing was dropped from the A2F-1 proposal. This is a 1959 model of one of these Concept 128 aircraft (Grumman)

about to unfold being somewhat unclear, since it was the subject of September 1956 engineering drawings and therefore pre-dates projects described below— did *not* feature the M-wing; it was an attack aircraft design to be powered by Allison T56 turboprop engines. It is mentioned here, a bit out of sequence with the flow of thought which follows because its was *the earliest date of any document I could locate and thus, to the purist historian, may be considered to mark the date when the A-6 story actually started.*

The Concept 128F aircraft was outwardly very similar to the Grumman OV-1 Mohawk (Design 134) which was being developed at the time for the US Army but which did not fly until 14 April 1959. Concept 128F, however, appears to have had overall much larger dimensions than the Mohawk. Since we shall see later that the J52 turbojet engine was decided upon very, very early during the events which followed, the 128F concept is, in a sense, unique.

Concept 128M seems to have been the very first of the most peculiar-looking M-winged aircraft and is the subject of an engineering drawing from Larry Mead's shop dated 27 February 1957. This seems to have been a more realistic approach to the TS 149 RFP, even though the official record of the competition begins a full month after 128M appeared on paper.

Concept 128M2 was to all appearances the second of Mead's M-winged attack aircraft, but seems to have been perhaps the first with the definitive configuration, including electronics stored aft of the crew compartment. It was sketched out by Mead on 14 March 1957.

Concept 128M3, which actually precedes the M-winged craft cited above, having been the subject of engineering drawings from Mead's shop dated 6 March 1957, had the main gear stowed under the twin engine nacelles, perhaps not a terribly comfortable arrangement, but one for which there seems to have been no alternative.

Concept 128M4, which followed on 16 March 1957 and was the subject of detailed engineering plans drawn by Mead, seems to have been little different from the others, but the notes accompanying the engineering drawings of this variant contain the first reference I could find to the J52 engine, which was to become, in fact, the powerplant for the Intruder. (The day before, 16 March, Grumman had made known its interest in designing an attack aircraft for the Navy's needs and in attending a conference on those needs, as will be seen.) Since the US Navy did not specify the powerplant to be used in the intended attack aircraft, it must be assumed that Grumman's engineers chose the powerplant they wanted long before coming anywhere near a decision on the shape of the airplane they were developing. Or, perhaps, it might be more accurate to say that the Mead design team had come to a preference for the J52 although, as will be seen, other possibilities at least reached the drawing board stage.

Concept 128N, also evolved during this period, featured a low wing and apparently did not have the M-wing configuration.

Concept 128P, which comes much, much later on 29 March 1959, long after the design and development activities described in the coming pages, seems to have been the first to actually resemble the A2F-1 (later A-6) Intruder which materialized as an actual aircraft.

Sequence of Events

So much, then, for the preliminary designs which are now history and which did not, themselves, lead to cutting metal. Far from merely saying that the US Navy needed a new attack craft because of Korean War lessons, let us now explore the sequence of events which produced the first A2F-1 Intruder.

This 1959 model of the yet-unfinished A2F-1 Intruder was made after the M-wing designs were dropped and the essential shape of the aircraft had been decided upon. The pointed nose was later redesigned and the Corvus missiles were dropped (Grumman)

On 27 February 1957, Mead was assigned as project engineer for the programme. According to Mead's own papers, the as-yet-unnamed aircraft dates to the summer of 1956 and the Navy's issuance of VA(X) operational requirement CA-01504, although Mead also notes that TS 149 did not arrive at Grumman until 4 March 1957. On 5 March 1957, in a letter to Grumman from the US Navy's Bureau of Aeronautics, signed by Rear Admiral W A Schoech, the Navy followed up on TS 149 by

Seen with the first A2F-1 Intruder are the all-important triumvurate who were present at the creation : Bruce Tuttle, head of the A2F-1 programme, Bob Smyth, test pilot, and Larry Mead, designer
(Grumman)

announcing its intention to conduct a competition for an 'all-weather, low-altitude attack weapons system'. It was fashionable in those days to talk about weapons systems rather than airplanes but it was indeed an airplane the Navy was looking for.

On 12 March 1957, while those M-winged drawings were taking shape at Bethpage, BuAer invited Grumman to attend a presentation on its intended 'all weather attack system'—the presentation having been developed by and to be held at Central Systems Laboratory of the University of Illinois at Monticello.

On 15 March 1957, again a date when the M-winged work was appearing on paper in the Bethpage design shop, William T Schwendeler, Grumman's senior vice president, responded affirmatively first to the Navy's invitation to submit a design and secondly to the invitation to the Monticello conference. It must have been a very interesting conference indeed. No record seems to have survived in any readily accessible location to show what was discussed. It can only be guessed that the seven other aircraft manufacturers who were eventually involved in one way or another were also in attendance.

A few days later, on 9 April 1957, contract design data requirements were transmitted to Grumman. A 1 May 1957 amendment to existing documents from BuAer included a provision for the new aircraft to carry drop tanks and provided details about the type of UHF radio to be used. On 3 May 1957, BuAer provided Grumman with information about the '20-mm aircraft gun Mk II'. Eventually, of course, the Intruder which never carried any gun at all remained in service long after the Mk II cannon was replaced and deemed obsolete.

Cost Proposal

These exchanges of pieces of paper were taking place at a time when Dwight Eisenhower was president, Elvis Presley was at the top of the charts with 'Heartbreak Hotel', the USS *Forrestal* (CVA-59) was entering the Fleet as the first of a new class of supercarriers, a potential war seemed to be brewing over the Formosa offshore islands of Quemoy and Matsu, and Marilyn Monroe was appearing in the brilliant motion picture 'Some Like It Hot'. Teenaged boys wore their hair in ducktails and went to school in black leather jackets. It was still possible at just about any Naval Air Station to spot an ageing F4U Corsair

or TBM Avenger. The personal aircraft of the Chief of Naval Operations was an R6D-1, or military DC-6. Grumman was busily manufacturing the F11F Tiger fighter and S2F Tracker anti-submarine airplane. A new automobile in the United States sold for $1,600, a four-bedroom house for $25,000. A new attack aircraft for the US Fleet was scarcely the first subject on most people's minds, but a few enthusiasts knew that if Grumman *did* manufacture a naval attack aircraft, it would be the company's second, the first having been the prop-driven AF Guardian anti-submarine craft. Grumman's second Navy attack craft, under the designation system then in use, would be the A2F-1.

On 15 May 1957, a letter from BuAer invited

The A2F-1 Intruder mockup, replete with real glass for the windshield takes shape at Grumman in 1959. This wooden replica bore a close resemblance to the actual airplane which followed
(Grumman)

Grumman to submit an engineering and cost proposal for the new attack aircraft. Various details of the proposal were addressed further in correspondence in June and July, including details about launching the new aircraft from a carrier by catapult and details about ordnance stores the new aircraft would carry. All of this correspondence was, in a sense, prelude.

A key milestone came on 16 August 1957. A letter from Grumman to the Chief, BuAer formally submitted the company's proposal for what was now called Grumman Design 128 (although it was a totally different aircraft than the series of M-winged concept 128 airplanes Mead had sketched earlier). The letter was signed by LeRoy R Grumman, chairman of the

The wooden mockup of the A2F-1 Intruder even included a starboard folding wing. The mockup was intended to familiarize Navy Bureau of Aeronautics people with the forthcoming design
(Grumman)

board, and was hand-carried to BuAer by Horace Moore on 19 August 1957.

Described in the letter as a 'sophisticated workhorse', the Grumman Design 128 would be a conventional sweptwing aircraft with two non-afterburning Pratt & Whitney J52 turbojets and with staggered side-by-side seating for its two-man crew. The letter had been completed in 'final' but was subsequently changed when someone—probably Mead rather than Grumman—pencilled-in last-minute alterations. For example, reference to STOL (short takeoff and landing) 'application of jet-thrust deflection' was not in the first 'final' version of the letter but was added to remind BuAer of the company's plan to include tilting exhaust nozzles in the new aircraft.

Grumman proposed to complete a mockup eight months after getting the go-ahead, to achieve a first flight 27 months following go-ahead, and to begin BIS trials (Board of Inspection and Survey) 42 months from go-ahead.

The competition must have been intense, for the eight competing companies submitted no fewer than

G-59492

GRUMMAN A-6 INTRUDER

twelve designs, four of them powered by turboprop engines. (Mead's design team had by now gotten far away from its early turboprop 'Navy Mohawk'.) Grumman's Design 128 was the only aircraft in the competition with side-by-side seating. The company's files indicate further correspondence with BuAer on 30 August 1957, apparently on cost and ordnance options.

Lest any assumptions be made about American technological superiority at this juncture, the times should be set in their proper context by mention of another event far from Grumman's offices: on 4 October 1957, the Soviet Union orbited Sputnik, the world's first space satellite. Larry Mead, Navy class desk officer Commander Bill Ditch, and Grumman people Bob Nafis, Gene Bonan and Dan Collins, among others, suddenly found themselves labouring in a setting where their country was confronted by an able and up-to-date adversary.

Five months after the letter bearing Grumman's signature, on 30 December 1957, BuAer reached its decision. Boeing, Douglas, Vought and Martin had submitted two designs each, one turboprop and one jet. Bell, Lockheed and North American had weighed in with one design each. Bell's was a V/STOL proposal and was quickly dropped. Turboprop and single-engine designs were also quickly dropped, leaving Douglas, Vought and Grumman the finalists. When the Navy's decision reached him in the form of a phone call—a long-distance call was still a major undertaking in those days—Mead celebrated by taking his wife and four children out for lobster dinner at Link's. On 2 January 1958, the US Navy announced that Grumman had won the competition. A document dated 10 January 1958 apparently contains the first reference to the A2F-1 designation, found in a Grumman request for $3,999,623 in pre-production costs. As work proceeded on the A2F-1, on 6 February 1958 BuAer informed Grumman by letter of its wish that the new attack aircraft be able to employ guided missiles. Grumman was tasked to analyse the Bullpup and Corvus missiles.

On 14 February 1958, the Navy issued cost-plus-fixed fee contract No 58-524C for $3,410,148 to cover a mockup and mockup photos of the A2F-1. The US Navy's formal intention to enter into a contract for the actual airplane was signified in a 29 April 1958 letter. Work on the new aircraft was well under way when, on 12 March 1958, Grumman's Bill Cochran was named programme manager for the new aircraft. This was to be a short, sad interval in the airplane's history for in May 1958, Cochran was killed in the crash of a B-52 bomber at Chicopee, Massachusetts. The following month, Bruce Tuttle was named programme manager.

Looking back through the clear lens of hindsight, it is interesting to note that the A2F-1 Intruder might have become a platform for the Corvus missile but didn't, might have later employed the Condor missile but didn't, and might have mounted an internal 20-mm gun—but did not. As for the Condor, Grumman engineer Joseph Le Strange remembers that the missile, never successful in operational services, was the subject of 'one of the world's longest development programmes'. Le Strange notes that this aspect of the Intruder's development lasted from 1962 to 1973. About a dozen Intruders eventually had wiring for the Condor, which used a TV sensor in the nose of the aircraft, but the missile simply never quite made the grade.

The possibility of a gun aboard the Intruder was apparently discarded early in the A2F-1 programme. Says Le Strange, 'There was no reasonable place in the configuration for it [a gun]'. 'We also reviewed gun pods and did some flight testing, but they were unreliable.' Le Strange also notes that early attempts to mount Sidewinder air-to-air missiles on the Intruder were made locally at NAS Oceana, Virginia, where a detachment from squadron VX-5 tried Sidewinders mounted on the side of the fuselage. This was another possible weapons store which never materialized on the operational aircraft.

Of the Bullpup missile—remembered by so many American airmen for bouncing off bridges in North Vietnam—Le Strange remembers that the A2F-1 Intruder design would have employed considerably more than the 'crude Bullpup arrangement with toggle and visual delivery' found on aircraft like the US Air Force's F-105D Thunderchief. The Bullpup was to be fitted to the A2F-1 Intruder in conjunction with an automated guidance system called BAGS.

On 3 February 1959, a letter to BuAer from Bruce Tuttle, Grumman's A2F-1 Programme Manager, provided a brochure on the A2F-1 airplane as then proposed. By 1959, the first A2F-1 Intruder was taking shape at Grumman's Calverton, Long Island facility about an hour's drive from the Bethpage headquarters. At the same time, engineers at the Grumman Aircraft Engineering Corporation—the firm's full name at the time, although it has since become the Grumman Corporation—were working on an Intruder which *would have appeared in Air Force markings*, had it been built.

Attributed primarily to designer G F ('Fritz') Dunmire who was on Mead's team, Grumman Design 128B was a response to Air Force requirement SR-195 which apparently never produced a finished airplane, perhaps because a change of administration and the outset of the Kennedy era was imminent. Design 128B was a tactical strike reconnaissance system employing the basic A2F-1 Intruder airframe but also carrying a ventral 'canoe' fairing which contained a recce pod with generator and constant-speed television viewer, AN/APQ-55 side-looking radar (the acronym SLAR was not yet in use), integral fuel, and KA-5 cameras. In addition, Design 128B was to carry Bullpup missiles under the wings. Externally identical to the actual A2F-1 Intruder, Design 128B exists today only in the form of a small scale model at Lovisolo's historical office, painted in Air Force markings. Since the Navy and Air Force still had separate designation systems at this time, and the 'A' for attack designator was *not* used by the Air Force, it can only be guessed what that service's version of the Intruder might have been called. Perhaps, given an 'F' for fighter appellation, it might have become the F-112A Intruder! But this was not to be. (This Dunmire design is mentioned again in this narrative's later section on Model numbers assigned to Intruders.)

Heaven only knows how many other possible Intruders failed to materialize. One which seems to have been contemporary with construction of the A2F-1 prototype—and with the Mohawk—was Design 128G-5, which employed the Intruder airframe for a US Army deep-penetration surveillance aircraft based on the A2F-1.

This 'Intruder that wasn't' would have deleted naval provisions and would have had an empty weight of 21,016 lb (9533 kg) carrying Westinghouse target locating and mapping system (ATLM), with pod-mounted SLAR antenna, KA-30 camera installation (similar to that on the Mohawk), AN/APN-118 Doppler navigation system, and AN/APG-53A terrain-clearance radar.

Like the Air Force Intruder, the Army Intruder was not to be. The Navy's A2F-1, however, was about to take shape at Bethpage.

Another milestone came on 26 March 1959 in the form of the US Navy's contract for A2F-1 development and for two batches of four airframes each, at a price of $101,701,000. The first eight airframes (bureau numbers 147864/867; 148615/618) were on the way and this gave the programme a certain inevitability. Working on an ambitious schedule, the design team sent its first detail drawings to the shop on 15 April 1959. Incredibly, the new aircraft would spread its wings only one year and four days later!

On 26 June 1959, Grumman reported to BuAer on efforts to reduce the vulnerability of the A2F-1 to infrared seeking missiles and proposed an Infrared Suppression Programme in connection with early developmental tests of the A2F-1.

Although STOL capability proved far less important than it seemed at the time, Grumman's attention to STOL characteristics may have been critical to the firm's winning the contract in the first place. Certainly, STOL was of great interest to BuAer during this period, Navy engineers looking at some seem-

Early model of the A2F-1 Intruder is painted in a camouflage scheme. The reason is unclear, since neither Air Force nor Navy aircraft were camouflaged in the late 1950s, but much later, in 1965, a few Intruders in the Vietnam combat zone did *wear paint schemes of this kind* (Grumman)

ingly wide-eyed proposals from Britain's Hawker-Siddeley firm. Design 128Q, as the A2F-1 was known to Grumman, was intended to meet the Marines' STOL requirement—clearing a 50-ft (15.24-m) obstacle with a takeoff run of 1,500 ft (457.2 m)—by employing variable geometry tailpipes with its J52 engines. The tailpipes were canted downward at 23 degrees and enabled a significant reduction in both takeoff speed and takeoff distance. In the end, the STOL tilting-tailpipe feature was installed only on the first seven A2F-1 prototypes and the Intruder achieved good short-field performance not through vectored thrust but through the use of air brakes although, as will be seen, these were far from trouble-free.

In natural metal and with standard Navy marking, the first A2F-1 (bureau number 147864) was rolled out in early 1960. It might help set the date in context to note that Eisenhower was president, John F Kennedy was beginning his quest for the democratic nomination, and, on the very day of the A2F-1's first flight, 19 April 1960, rioting students in Korea began the week of upheavals which would throw President Syngman Rhee out of office. Incredible as it seems,

today's standard US Navy medium attack aircraft was coming into the world while Ike and Rhee were in the headlines!

On 1 April 1960, the A2F-1 prototype was outdoors for the first time, for aircraft fuel systems tests conducted at Plant 4 at Bethpage. The A2F-1 was then slow-taxied in tests at Bethpage, and put aboard a truck to travel to Calverton by land. It would be nice if there were something dramatic to impart about the first flight of the first A2F-1, piloted by Bob Smyth (who has since become an executive of the Gulfstream Aerospace Corporation, itself once a part of Grumman but no longer). In fact, little drama seems to have attended the A2F-1's maiden voyage. Le Strange, who was present for that event, remembers how A2F-1 number one had been assembled at Plant 5 (at Bethpage) 'with an army of guys', the suggestion being that in a later era, production methods were less cluttered. Of the first flight itself, Le Strange recalls that the chase plane did not stay with Smyth. 'The first flight was made with the [vectored thrust] pipes down, the gear down, and everything hanging out.' Smyth took off from Calverton and flew the short distance to Bethpage, so that employees at

The first A2F-1 Intruder, bureau number 147864. After being extensively modified, on 11 July 1962 the aircraft now has wingtip speed brakes which eventually became standard on all attack variants, and a new vertical tail shape adopted for all production machines. The hard-working first airframe was being used for spin tests, the cylindrical device at the tail being a spin parachute
(Grumman)

the firm's headquarters could have a glimpse at the result of their labours.

Smyth then landed routinely at Calverton. The first flight was deemed successful and Grumman executive Roger Kahn sent a cable to James O'Grady, the manufacturer's field service representative at NAS Oceana, Virginia (future home of the east coast Intruder community) telling him so. Today, when direct-dialing is so much in vogue, the preserved original of that Western Union telegram is yet another reminder of just how long ago the Intruder came into the world.

Other Proposals

In January 1959, the A2F-1 acquired an additional two feet of wingspan and an additional 1,000 lb (454 kg) of fuel volume. Mead's design team had found an error in the cruise drag calculations for the new aircraft, the result of the wrong figure being used to relate actual wing area to that of a wind tunnel model. By making the change in the wing dimensions, Mead's team improved aspect ratio from 5 to 5.31 and found space for the additional fuel, thus retaining the

intended operational radius of the A2F-1. Negotiations for this minor change in configuration were handled smoothly over the phone between Mead and BuAer class desk officer Lieutenant Commander C P (Bud) Ekas.

Empty Shell

As yet, of course, the A2F-1 was only an airframe. 147864 would not have been capable of dropping a bomb on anybody. Nor would it have navigated very far, since no radar was installed beneath the blunt nose. Nor was radar installed aboard the number two A2F-1 (bureau number 147865), which flew on 28 July 1960. Not until the fourth airframe would the Intruder acquire radar and (another important feature) an air refuelling probe.

Speaking of the number two airframe, which made its initial flight on a hop from Bethpage to Calverton, this A2F-1 had a controllable fuel shut-off valve between the two main fuel tanks in the fuselage, to enable engineers to vary the CG (centre of gravity) in flight tests. Due to what Mead calls 'a nomenclature problem', this valve was closed instead of open during 147865's maiden flight and the rear tank, which was the engine feed tank, went dry. Both engines flamed out just as the A2F-1 entered the downwind leg for landing at Calverton. Pilot Ernie Vonder Heyden made an unplanned but completely successful dead-stick landing, demonstrating, as Mead says it, 'the good flying qualities [of the A2F-1] and an honest airplane in the process'.

On a subsequent test flight, Vonder Heyden was checking out the fuselage side speed brakes, with power on. He noted a sluggish response. Mead: 'When we checked the records after flight we found that the aerodynamic hinge movements on the slab tail exceeded the hydraulic actuator capacity and the actuator stalled for a few seconds, fortunately with no disastrous results.' It was this situation which resulted by January 1961 in the A2F-1 stabilizer being moved aft 16 inches (0.4 m) beginning with airframe number three (bureau number 147866).

If the Intruder's navigation and weapons delivery system (NWDS) was to introduce a revolution, the basic design of the Grumman airplane was conventional enough. Fuselage construction of the largely aluminium A2F-1 was of the semi-monocoque type with the exception of the lower half, where a structural keel beam of steel and titanium was employed between the engines, the nonstructural doors surrounding the engine bay. The A2F-1 held its two 8,500-lb (3856-kg) thrust Pratt & Whitney J52-P-6 turbojets in what might now be called conformal pods, set low under the fuselage to leave the central structure free for the capacious fuel tanks that gave aircraft the combat radius required by TS 149.

Apart from the STOL jetpipes that were soon dispensed with, to reduce approach and touchdown speeds and to achieve the highest possible lift coefficient the wing of the A2F-1 was designed with very little sweep (25 degrees at the quarter-chord). The wing consisted of five major sub-assemblies, these being the centre section and port and starboard inner and outer sections. Inner wing panels were joined to a continuous box-beam which ran through the centre section by bolts while the outer panels were affixed at the wing-fold joint by four steel hinges. Wing-locking was achieved by hydraulically driven pins. Trailing edge flaps extended across an unusually high percentage of the 53-ft (16.15-m) wing span, eliminating the possibility of conventional ailerons. Lateral control was therefore provided by spoilers in

Ooops! The number six A2F-1 Intruder, bureau number 148616, seems to have suffered a faulty landing or, at least, a nose gear collapse. The external covering for the radome has begun to delaminate into tattered shreds. This incident seems to have occurred in about mid-1961 (Grumman)

the upper wing surface. These could be used to 'dump' lift once the aircraft had landed, thus increasing the load on the undercarriage and the effectiveness of the wheel brakes.

Without radar, of course, the Intruder would not do any navigating *or* any bombing. With presidential candidates Richard Nixon and Jack Kennedy spending much of 1960 each trying to prove that he was more likely than the other to defend Quemoy and Matsu, both navigation and bombardment were soon to become of importance not near Formosa but farther south, in a country few Americans had yet heard of. The bulbous shape of the A2F-1's nose, already mentioned, was dictated by the radars and NWDS needed to carry out the Grumman airplane's night- and all-weather attack mission.

The Intruder was designed for two radar antennas beneath its single, large radome—for the distinct search and track missions. The AN/APQ-92 search radar and AN/APQ-88 track radar (the latter ultimately replaced by AN/APQ-112) were advanced for the time. Yet these were only two components in a remarkable overall package called DIANE (digital, integrated attack and navigation equipment) which provided not merely fully automatic navigation and weapons delivery but, more, computerized handling of every aspect of the attack mission, overseen by the Intruder's second crewman, the bombardier-navigator (BN). (Diane was also the name of Grumman engineer Bob Nafis' daughter.) The DIANE system also encompassed Litton AN/APQ-61 ballistics computer, AN/ASN-31 inertial platform, CP-729A air data computer, AN/AVA-1 vertical display, AN/APN-141 radar altimeter, AN/APN-153 Doppler navigation, AN/ASQ-57 integrated electronic control and AN/AIC-4 crew intercom system (ICS), plus IFF, TACAN and ADF.

Smyth's tadpole-shaped prototype 147864 may have seemed an odd-looking bird to a generation of aeronauts weaned on the idea of flying higher and higher, faster and faster. The A2F-1 was most decidedly *not* conceived to soar to new heights and was not going to be faster than much of anything. Still, other features of the basic design illustrated how very practical the A2F-1 design was: the crew sat side-by-side in slightly staggered position (the pilot a few inches forward of the BN) in a cockpit which was spacious and well-forward with good visibility. The in-flight refuelling probe was located just ahead of the windscreen. No one wants loose ladders lying around an aircraft carrier's deck, so the Intruder provided its crew with integral boarding steps. Afterburning was deemed unnecessary for the subsonic strike role. A twin-nosewheel arrangement was ideally suited for carrier-borne nose gear tow.

Eventually, the prototype A2F-1 was fitted with aft-fuselage dive brakes also found on early production airplanes. It was discovered that these interrupted the airflow over the tail surfaces and the dive brakes on production Intruders were shifted to the wing tips.

The number ten A2F-1 Intruder, bureau number 149176, with the 'old' fuselage air brakes but 'new' vertical tail shape, sits on the ramp with a full load (Grumman)

LEFT
A-6A Intruder bureau number 149482, at NATC Patuxent River, Maryland on 11 June 1970. The tail marking indicates the Strike Test Directorate at 'Pax.' The Intruder's well-known 'birdcage' (ventral equipment package) is extended in this view
(Joseph G Handelman, DDS)

BELOW
Early in the development programme, Grumman and the Navy set forth to show that the Intruder could carry a payload. This view of A2F-1 Intruder number four, bureau number 147867, shows the aircraft carrying no fewer than thirty 500-lb (227-kg) bombs
(Grumman)

The final test for a new aircraft type : carrier qualifications.
A-6A Intruder No 8, with a mate waiting on the dock behind
it, is lifted aboard ship for its carrier trials. Second view
(right) shows the aircraft operating aboard the carrier.
December 1962
(Grumman)

Testing

Aircraft number four (bureau number 147867) seems to have been delivered with full onboard systems and was soon photographed toting a full load of bombs. On 18 September 1962, when the US forces adopted a new, joint system for designating their airplanes, it was the fourth airframe which was immediately repainted with the new name by which the A2F-1 would become known—A-6A. It was the eighth Intruder (bureau number 148618) which first went on shipboard in December 1962 to begin the type's carquals (carrier qualification trials) aboard USS *Enterprise* (CVAN-65).

In fact, formal Navy trials were conducted over a 14-month period, beginning in September 1961 with the first Naval Preliminary Evaluation (NPE-1), the carquals being a part of the successful NPE-2 phase which followed.

The final hurdle to be cleared before the A-6A Intruder's entry into service was the Board of Inspection and Survey (BIS) trials held beginning in November 1962 at the Naval Air Test Center, NAS Patuxent River, Maryland. Several airframes were

Illustrated on the ground on page 24, bureau number 147867 climbs above the clouds with a maximum load of thirty 500-lb (227-kg) bombs. It is doubtful whether this awesome bomb-load with its considerable built-in headwind was ever lifted for anything but the shortest missions, and a catapult launch in this configuration would have been interesting (Grumman)

assigned to the BIS exercise (reached well ahead of schedule) in order to cover all aspects of future Intruder use ranging from ground handling to weapons launching.

East Coast RAG

Fiscal year 1962 funding had moved beyond the first two batches of four A2F-1 Intruder prototypes each and had included a $150.3 million appropriation for a further 24 A-6As, this being followed in fiscal year 1963 (beginning 1 July 1962) with a further $159.3 million allocation for an additional 43 A-6As. Development was also proceeding with an electronic

*The US Marine Corps took an early interest in the Intruder.
A-6A aircraft 151567, coded DT-9, of the 'Bats' of
VMA(AW)-242 flies with full bomb-load near MCAS
Cherry Point in about 1965. This later became the first
Marine squadron in combat, arriving in Da Nang on 1
November 1966*
(USMC)

warfare version originally designated A2F-1Q (*not*, as widely reported, A2F-1H) and soon redesignated EA-6A. By the end of calendar year 1962, no fewer than twenty A-6As had been completed and from the 16th (bureau number 149482), the area of the rudder was increased to assist in spin recovery. (After installation of an extended drag chute apparatus, spin tests had been conducted using the first airframe.)

The first step towards introduction of a combat aircraft into US Navy service is the formation of a training squadron, known as a replacement (or replenishment) air group, or RAG. With the A-6 Intruder, as with most major US Navy types, there would eventually be two RAGs—one training squadron on the east coast to fulfil the needs of the Atlantic Fleet and another on the west coast for the Pacific Fleet.

The east coast Intruder RAG, it was quickly decided, would be the 'Green Pawns' of attack squadron VA-42, located at NAS Oceana, near Virginia Beach, not far from Norfolk where Atlantic Fleet carriers come to rest. first A-6A deliveries to VA-42 took place on 7 February 1963. Vice Admiral Frank O'Breirne, commander of US Forces Atlantic, officiated.

Years of effort and talent had produced an operational warplane the real significance of which was, as yet, far from fully understood. In the early days at Oceana, the A-6A Intruder was more an airframe than a system—as will be noted shortly, the true development of its internal systems would take years—but a solid beginning had unquestionably been made. In early 1963, a handful of Americans were advising—and fighting—in a distant corner of the world but most people in the United States, including most people learning to fly the Grumman A-6A Intruder, did not yet know about Vietnam.

Chapter 2
Combat: Vietnam

At the height of 1965–68 operations against North Vietnam, known by the hopeful codename Rolling Thunder, a press release from the US Navy's JOSN David W Butler described the A-6A Intruder in vivid terms:

'Close to the ground, a plane hurtles through monsoon rains to a target in North Vietnam. Inside the cockpit, the pilot watches a screen in front of him.

'A panorama of plains, rivers and mountains passes by and it looks like a cartoon on TV. Superimposed on the screen, a road curves through a canyon ahead and suddenly loops up above the horizon.

'The plane heads for the target, a square on the screen's horizon. Blinking red words appear on the console. "Target area . . . Attack . . ."'

The mid-1960s were a dramatic period of rapid movement in the A-6 Intruder programme, but the results were not always the Luke Skywalker drama described above. Once the 'Green Pawns' of VA-42 at NAS Oceana, Virginia, were in operation as the east coast/Atlantic Fleet RAG, the 'Sunday Punchers' of VA-75 began to receive the Intruder—perhaps a bit hurriedly. There were not enough airframes to bring the squadron up to full strength immediately. Worse, DIANE was not being a lady, at least not always. There were troubles with maintenance and reliability links in the all-weather navigation and weapons delivery system and more Intruders were 'down' more often than anybody wanted. It was also discovered that the Intruder (and the E-2A Hawkeye, introduced at the same time) required far more

During workups in the Caribbean Sea for what will become a combat cruise only weeks later, an Intruder is launched from USS Independence *(CVA-62) on 2 March 1965. A-6A aircraft 151583, coded AG-506, belongs to squadron VA-75 and has the black radome characteristic of early examples of Grumman attack craft*
(USN/PHC4 R C Lister)

support equipment than had been planned when the Navy's carriers were built, with the result that sailors on *Independence* and later *Kitty Hawk* had to scrounge mercilessly for space for checkout monitoring gear and other items. To some extent, these problems were also encountered by the 'Bats' of VMA(AW)-242, the first Marine Corps squadron to take the Intruder into battle.

Baptism of Fire

Although an east coast squadron, VA-75 intended from the beginning as the first Intruder combat squadron. The Sunday Punchers flew their first combat missions from the deck of USS *Independence* (CVA-62) on 1 July 1965. Early strikes were carried out against key highway bridges at Bac Bang and other targets in an area 80–125 miles (129–201 km) south of Hanoi. In these 'early days', MiGs were only

beginning to appear in force and North Vietnam was still building its formidable network of AAA (anti-aircraft artillery) emplacements and SAM missile batteries—but Ho Chi Minh's homeland was already a decidedly dangerous place to be. Added anxiety over DIANE's reliability was a very real part of life but was not anything *Indy*'s naval aviators needed at this time.

Early pictures of the Intruder can be recognized at a glance by the black radome on the aircraft (later, and throughout most of the airplane's career, these were white). Despite early difficulties, by the end of their first month in action a pair of A-6As were able to demonstrate one facet of their capability, depositing many tons of 500-lb (227-kg) Mk 83 and 1,000-lb (454-kg) Mk 84 bombs on the notorious Thanh Hoa power plant 80 miles (129 km) south of Hanoi. The attack, conducted under cover of night, employed radar to locate and identify the target complex. For

31

VA-75 may have been rushed into combat too quickly. These photos depict two of the squadron's original batch of A-6A Intruders during workups aboard USS Independence (CVA-62) in the Caribbean Sea in February 1965. In the first view, (above), aircraft 151581, coded AG-504, is launched just behind a waiting A-4E Skyhawk. In the second view, aircraft 151583, side number AG-506, is ready to go from the carrier's steam catapult (USN/PHC R C Lister)

the first time, rumblings from the Intruder began to suggest that the enemy would no longer rule the night. North Vietnamese industrial and transportation centres which had previously enjoyed respite from bombing during the nocturnal hours could now expect to be hit around the clock.

Or, at least, *maybe* North Vietnam would come to grief under the wings of the Intruder—but in early operations, systems reliability of the A-6A was an unimpressive 35 per cent! Part of the problem was a need for improvements to the search and track radars—quickly accomplished with retrofits—but the situation was exacerbated by the fact that coordinates of some of the early maps in use in Vietnam were as much as 3–4 miles (5–7 km) in error. Since the Intruder's system could only use the data fed into it, accurate geographic information was a 'must'. During the 1965–67 period, the Navy addressed this

issue by conducting an extensive aerial mapping survey of North Vietnam with the RA-5C Vigilante aircraft.

VA-75 struggled with the initially low reliability of the DIANE system and maintained its efforts against North Vietnam's highways, railroad depots, barracks, ammunition and fuel concentrations. The squadron was also conducting limited operations against communist targets in Laos, a fact which was known, it seemed, to everybody but the American public. The first recorded combat loss of an Intruder occurred on 14 July 1965 when aircraft 151584 went down in Laos. The following day, 151577 went down deep in North Vietnam. On 24 July 1965, 151585 was downed, also in Laos. It is not known whether any of the crews were recovered but, given the 'lone wolf' nature of many A-6A missions it is possible that they were not. If three losses in one month were not

enough—and they should have been, for it was not uncommon for a squadron using other types of aircraft to complete an entire cruise without losing a man or an airplane—VA-75 experienced its fourth loss on 17 September 1965, when 151588 was hit over North Vietnam and crashed at sea.

Surprisingly, according to a study by *Aviation Week and Space Technology*, Intruder losses were light during night operations, the very time when greatest danger would seem to be present. These averaged about one loss per 1,000 hours of combat flying. The loss of Intruders during daytime missions bordered on the appalling, however, and the first three losses of the war were attributed to a problem which was to continue to plague the attack community—premature detonation of their own bombs after release.

This happened when the Intruders were still carry-

Seen at about the time the squadron 'shipped over' to begin combat operations in Southeast Asia, these formation shots show five of VA-75's early, bomb-laden A-6A Intruders in flight. The squadron went to war wearing its 'east coast' AG tail code
(Grumman)

ing primarily World War 2 ordnance, including a mechanical nose arming fuse, on the A-6A's early multiple bomb racks (MBRs). When the aircraft released its bombload at a steep angle of dive, the fuze would become armed approximately 0.7 seconds after release. This interval did not allow time for the bomb to be anywhere but directly below the Intruder. The MBR did not in any way propel or boost the bomb at the time of release, so in the first three loss incidents the bombs simply tumbled into each other and exploded. Eventually, the Navy had to limit the Intruder to straight and level drops until it could retrofit the Douglas-built multiple ejector racks (MER) and triple ejector racks (TER) which used an explosive charge to release the bombs, virtually eliminating the cause of the problem. It must be emphasized that the problem with bombs and bomb fuzes was *not* unique to the Grumman Intruder; indeed, the Air Force F-4D Phantom and Navy A-7A

Corsair, among others, suffered undue losses for the same reason.

The first combat cruise of the A-6A was very much a story of mixed results, and VA-75 and *Independence* retired from the combat zone in November 1965. The losses must have hung heavily in the men's minds. They had inflicted some damage on the enemy, but perhaps not as much as they'd hoped. If nothing else, they had made the Air Force conscious of the night- and bad-weather mission. Following a Southeast Asia visit by PACAF chief General John Ryan, a special unit of F-105F Thunderchief was put together at Korat, Thailand, to assault North Vietnamese targets during the nocturnal hours and in times of foul weather. Ryan's Raiders, headed up by Lieutenant Colonel James McInerney, performed valiantly but had to face the fact that, unlike the Intruder, the Thunderchief had not been designed for this mission. Years later, McInerney, who retired

at lieutenant general rank, spoke to the author about the A-6 Intruder with the kind of admiration an officer does not bestow easily when the airplane belongs to a rival service.

In April 1965, a second Atlantic Fleet squadron, the 'Black Falcons' of VA-65, transitioned from the A-1H Skyraider to the tadpole-shaped Grumman. A part of carrier air wing eleven (CVW-11), in September the squadron deployed to the Tonkin Gulf and flew in combat over both South and North Vietnam. There are rumours that a VA-85 crew tangled briefly and inconclusively with a MiG-17 (the Intruder having no gun or missile to fight with), a happening which must have caused some adrenalin to churn.

VA-85 went into the combat zone abroad USS *Kitty Hawk* (CVA-63) with the squadron headed by Commander J E Kelley, a tough-minded and demanding skipper who got results because he was usually out in the front when it was time to attack the difficult

The 'Green Pawns' were the east coast Replenishment Air Group (RAG) for the Intruder community and trained the first crews to take the aircraft into combat. On 25 August 1963, A-6A number 149941, side number AD-504, participates in early carrier operations in the Atlantic aboard USS Forrestal (CVA-59) (USN)

LEFT
The original A-6A Intruder remained in service long after the E model arrived. The 'Swordsmen' of VA-145, seen operating from the deck of USS Ranger (CV-61), was still flying the A-6A on 1 October 1975. Aircraft 155669, coded NE-505, carries colourful markings of a kind no longer found on any carrier-based aircraft (USN)

targets. Like the squadron which preceded them, Kelley's men had accelerated the A-6A training schedule in order to meet Kitty Hawk's Westpac (Western Pacific) deployment date. VA-85, at least, had A-6A Intruders which were somewhat improved, having for the first time a radar warning receiver (RWR) as well as some improvements in the search and track and terrain-clearance radars.

A massive land battle was the scene of early action by VA-85. Following the near-slaughter of South Vietnam's 7th Regiment in a prolonged, close-quarters battle, VA-85 bombed and rocketed Viet Cong forces in the Michelin Plantation, only 45 miles (72 km) north of Saigon. This was not the highly specialized mission for which the Intruder had been designed, and at times it exposed aircrews to small-arms fire against which they were vulnerable, but the results were a testimony to the Grumman airplane's flexibility. The Viet Cong took heavy casualties and relief soon reached friendly ground forces.

Losses continued to be high, however. On 21 December 1965, Intruder 151781 went down over North Vietnam. 151797 followed on 18 February 1966. On 17 April 1966, aircraft 151794 was hit over North Vietnam and crashed at sea. Five days later, 151798 went down over the north. On 22 April 1966, Intruder 151785 was hit up north and crashed at sea. On 27 April 1966, 151788 suffered a similar fate. In its prolonged and admittedly exceedingly difficult cruise, VA-85 had lost no fewer than *six* airplanes, possibly a record for any cruise during the entire conflict.

Because a need for some improvements was obvious, the Navy embarked on a system improvement programme (SIP), beginning in the summer of 1966, for the Intruder's search and track radars—the Norden AN/APQ-92 and AN/APQ-112 respectively. The latter had been an improvement over the AN/APQ-88 designed and built by the Naval Avionics Facility at Indianapolis, Indiana, and installed in the earliest Intruders. The equipment was still essentially 1956-era technology. The SIP for the '92' included the use of new crystals to obtain better mean times between failure (MTBF), employment of an internal Klystron tube as a substitute for an external microwave source to reduce corrosion, and other minor steps. The SIP effort contributed to much improvement in the Intruder's reputation and in the morale of the men flying it.

Widening War

As the number of Intruders in the Navy and Marine inventory grew, additional squadrons began to work up although, contrary to the Navy's usual two-ocean policy, the Intruder effort remained concentrated on the east coast. A second Marine Corps A-6A Intruder squadron was deployed to Vietnam when VMA-533's 'Hawks' set up shop at Chu Lai, South Vietnam. The electronic-warfare EA-6A Intruder

was also coming into service with the Marine Corps and the east coast's Marine Composite Reconnaissance Squadron Two (VMCJ-2) operated briefly at Da Nang before being replaced by VMCJ-1, which otherwise would have been home-ported at MCAS Iwakuni, Japan. Another Navy squadron, the 'Boomers' of VA-165 entered combat in December 1967 from the decks of USS *Ranger* (CVA-61), VA-75 having begun its second combat cruise the previous month after shifting to *Kitty Hawk*. In January 1968, with the communists' Tet Offensive capturing headlines and stunning American TV viewers, the 'Black Panthers' of VA-35 arrived in the battle zone aboard the nuclear-powered USS *Enterprise* (CVAN-65). President Johnson's Rolling Thunder campaign against North Vietnam was rushing to its climax, and the defences around the Hanoi region were now the most formidable ever seen.

By late 1966, the 'Tigers' of VA-65 commanded by Commander Robert C Mandeville were in the battle zone aboard USS *Constellation* (CVA-64). *Time* magazine interviewed Mandeville, looked at DIANE which it called 'a spaghetti bowl of instruments that combines . . . radars, an inertial navigation unit, and a small computer', and pointed out that the A-6A Intruder really proved itself in the October 1966

Many A-6A were converted to become KA-6D tankers. These retain the fuselage airbrakes which were moved to the wingtips of most other Intruder variants. KA-6D aircraft 151591, side number NL-506 of the 'Knight Riders' of VA-52 aboard USS Kitty Hawk *(CVA-63) accompanies a fellow KA-6D over Southern California on 23 March 1979. Mark 82 500-lb (227-kg) bombs are not an unusual payload for the tanker version, which is dual-capable to maximize the striking power of Intruder squadrons (USN/PHC Arthur E Legare)*

RIGHT
Intruder and intruder. Over the Pacific Ocean in about 1980, an A-6A Intruder, bureau number 155670, side number NL-503, plays escort. The Soviet Ilyushin Il-38 May maritime patrol aircraft appears to be snooping into the activities of the US carrier battle group in the area (USN)

period when the northeast monsoon brought its annual drenching rains and galeforce winds. While Phantoms and Skyhawks were grounded, Mandeville's squadron flew 40 per cent of the missions logged by all of the squadrons abroad the three carriers operating in the Gulf of Tonkin.

Flying the Intruder

What was it like to live aboard a pitching, growling aircraft carrier and to fly the revolutionary but controversial A-6A Intruder in combat? A 'mission profile' for the Intruder was put together based on comments by aircrew members. To fly the mission with them, this narrative now reverts to the present tense.

Ike Claney and Tim Parker end their Wednesday night not yet knowing which target they'll be assigned on a nocturnal penetration of Ho Chi Minh's airspace tomorrow night. Both men are 28-year-old lieutenants, their insignia of rank the twin bars known as railroad tracks. Claney is a pilot, Parker a bombardier-navigator. They consider themselves among the best-trained and most motivated people in naval aviation. The A-6A Intruder community is small and close-knit, admittedly lacking some of the flair and razzle-dazzle of what Parker contemptuously calls 'the fighter guys'. Their community is a small, tight clique of good men with a special pride in their mission.

Claney and Parker are almost certainly aware that the Intruder has been taking its losses in operations against North Vietnam. The 'Tigers' of VA-65, operating from USS *Constellation* (CVA-64) lost 151816 when it was hit over North Vietnam and crashed at sea on 25 June 1966. This was followed by the loss of 151822 over North Vietnam on 27 August 1966. Now that Claney and Parker are with VA-85 for its second combat cruise on *Kitty Hawk*, their squadron has already lost 151590 over North Vietnam on 1 January 1967. Some of the crews have

been rescued, of course, but in general an Intruder crew does not stand a good chance of being saved because they operate deep inside North Vietnam where SAR (search and rescue) forces have yet to become very effective.

Our pilot and BN belong to the sole medium-attack squadron found on most US Navy carriers. Each squadron has 16 crews and is supposed to have 14 aircraft, but in Southeast Asia they are operating with a scarcity of Intruders and VA-85 may have as few as nine airframes. Grumman is developing a KA-6D tanker version of the airplane, and eventually four of the 14 airplanes in a squadron will be tankers, but in early 1967 the Intruder receives fuel in mid-air from the Douglas KA-3B Skywarrior.

On our typical mission, the A-6A Intruder crew will strike a North Vietnamese trestle bridge located 240 miles (386 km) from *Kitty Hawk* and 140 miles (225 km) inland.

Wednesday: 2200 hours

Kitty Hawk is a brooding silhouette plowing through sea 100 miles (161 km) from the enemy coast. 24 hours before the mission, an Air Plan is put out by the carrier's staff, with major input from the CAG (carrier air wing commander). Itself a very general description of the ship's operations for the following day, the Air Plan goes quickly to the squadron's scheduling officer who, in turn, takes a hard look at how many of VA-85's people and airframes will be used in the planned bridge strike. At this juncture, the carrier's Strike Operations people consult with

the 'Black Falcons' operations officer about the ordnance to be carried. At this early juncture, the ordnance load is decided upon: the A-6A will carry a centreline Aero 1-D droptank containing 300 US gal (2,000 lb) of fuel and twelve 500-lb (227-kg) Mk 82 bombs with Mk 904 mechanical nose fuzes, the bombs being divided into clusters of six each on MERS outboard at ordnance stations one and five. This is a typical but moderate payload for the Intruder and one which should permit easy ingress to the target area and, if needed, loiter time over target. Air refuelling will be used only on the way home and only to provide an extra margin of safety.

The weatherman is calling for low to medium cloud cover with a 40 per cent chance of thunderstorms. This is no joke—flying hail can pierce an airplane's thin metal skin like machine gun bullets—but while this warning might cancel a Skyhawk or Corsair mission, it is merely an incidental item for prudent caution by the Intruder men.

Before going to sleep on the night before, A-6A Intruder crews slated for the mission assemble in CVIC (the carrier's intelligence centre). Tonight, *two* Intruder crews meet. Parker and the other BN plan the navigation. They sit back and scutinize the obvious: where is the target? Where will *Kitty Hawk* be when they're ready to recover? They select a coast-in point so that they can cross the beach at a readily-definable landmark which shows well on radar or visually, such as a peninsula with a hook of islands. The two Intruder pilots decide that they will cross the beach together but will vary their routes slightly, one A-6A continuing straight ahead, the other turning 30 degrees to arrive at the target 90 seconds later than the leader (about 12 miles apart). The men will fly to their target at an altitude of 500 ft (152 m) at night.

Still during the night before the mission ('Intruder guys do it at night', reads the bumper sticker distributed by the squadron coffee mess officer), the A-6A pilots do their weaponeering. They determine settings on the A-6A armament panel. They confirm both stations, the method of bombing—they'll drop the bombs in a string at an angle—and review the attack profile to fit safety standards. Almost all of this skull-session decision-making is performed by the aircrews alone. The Navy, especially the Intruder community, places strong importance on personal judgement and initiative.

Thursday: 0800 hours

Lead of the two-plane division, Ike Claney has the callsign ROMPER 502. The men awaken at 8 or 9 am. They go jogging on deck to release energy. A light drizzle warns of worsening weather. During the morning, Ike and Tom talk with the crew of the E-1B or E-2A early warning craft which will not cross the beach with them but will remain off the coast and delouse them on the way out (that is, make certain they

are not being tailed by MiGs). They also talk with the carrier's F-4 Phantom pilots. At this time, another crew is flying the same A-6A Intruder airframe on the first mission of the day, but our two crews enjoy a late lunch at 1400 and a quick nap at 1500.

Thursday: 1830 hours

Intruder, Phantom and Tracer crews gather for the Air Wing brief. This is held in VA-85's squadron ready room. This discussion of the coming mission is followed by an Intel (intelligence) brief on closed-circuit TV, providing the Intruder crews with information about the enemy's defences. After the Intel brief, the two A-6A crews go through a standard briefing checklist covering everything they'll need to know.

Finally, in the last 5 or 10 minutes, each individual crew briefs itself. Claney and Parker can understand each other almost by intuition but they rigorously go over every detail out loud to avoid misunderstanding. Neither man forgets that in an abrupt crisis, they may be forced to resort to their Martin-Baker Mk GRU-5 ejection seats. Ike Claney says again, as he does before every flight, that if the crunch comes and he wants to order an ejection he will say three times, 'Eject, eject, *eject*!' He'll leave the airplane simultaneously with the third utterance. If the ICS (intercom system) is not working, he'll use hand signals.

It is now 2100 hours, one hour before launch. The flight crews go to Maintenance Control. Here, they read the discrepancy books (complaints about their individual A-6A over the past ten days; a characteristic Intruder problem is a malfunctioning electrical system). As pilot, Ike Claney signs the 'A Sheet', a small sheet of paper in the plane captain's hands which signifies that the A-6A is fully ready. If appropriate, the crew gets a verbal debrief from the crew which has just finished using the airplane.

Now, Ike and Tim go to the paraloft to obtain flight gear. It is cramped here, so pilot and BN hurriedly gather G-suit, torso harness, survival gear, PRC-90 survival radio, and helmet. The APH-1 helmet, or 'brain bucket' is a heavy item, both whilst being carried and while being worn.

En route to the airplane, Tim Parker stops at flight deck control (forward of the island on *Kitty Hawk*'s deck) to drop off a weight sheet. This contains fuel and ordnance weights for the catapult officer responsible for launch. A clerical error here, and an improperly-set steam catapult will send off the Intruder with just enough acceleration to plop it into the drink.

The crew arrives at the plane and is greeted by the plane captain—one of the unsung heroes of naval aviation—who views Airplane 502 as his personal 'baby'. Having readied-up the A-6A with loving care, the plane captain now stows gear inside the aircraft. Each crewman gets into the A-6A via his own internal boarding ladder. The men clamber down into their seats and the plane captain helps them to strap in. Power comes to the Intruder from deck cables. *Kitty Hawk*'s Air Boss (head of the carrier's air department) is in charge of the flight deck crews and orders

Intruder and lifeguard. On a Mediterranean cruise aboard USS Independence *(CV-62) in early 1975, an Intruder is depicted with an SH-3D Sea King 'plane guard' helicopter in the background. Intruder's nosewheel is fixed to the bridle for launch from the carrier's catapult. Launching an aircraft from its nose gear is commonplace today, but the Intruder was the first aircraft with this feature* (USN)

engine starts. A sound peculiar to the Intruder, a sigh, precedes a heavy whine as the twin 9,300-lb (4128-kg) thrust Pratt & Whitney J52-P-8A engines kick into life. The crew closes the canopy and runs through a pre-takeoff checklist on ICS. A yellow-shirted plane director hand-signals to indicate that they are ready to move, takes their tiedown chains and chocks, and gestures the pilot to taxi.

Thursday: 2200 hours

A-6A Intruder pilot Ike Claney taxies to his catapult. He taxies into the shuttle, so that a towbar link fixes the airplane to the catapult. The Intruder introduced the method of launching an aircraft via its nose wheel,

Squadron VMCJ-1 introduced the EA-6A Electric Intruder to combat in markings like those shown on aircraft 151598, side number RM-7, seen at MCAS Iwakuni, Japan during the Southeast Asia conflict. In background are RF-4B Phantoms also operated by the squadron at the time (USMC)

LEFT
In its final period as a user of the EA-6A, from September 1974 until September 1975, squadron VMCJ-1 operated from MCAS Iwakuni, Japan and used the markings seen here. During the latter month, all EA-6A resources in the Marine Corps were consolidated into VMAQ-2 at MCAS Cherry Point, North Carolina. Meanwhile, the three-digit side number RM-614 on aircraft 156984 indicates its readiness to operate aboard ship (USMC)

revolutionary at the time but later introduced throughout the Fleet. The deck crew scrutinizes the Intruder as it strains to go, looks for oil leaks, open panels, improper control surfaces. Only after this, the 'yellow shirt' passes the A-6A to the catapult officer. Since darkness has now fallen, Ike Claney signifies readiness by turning the Intruder's lights on.

At 2200 hours, several tons of brute force in the form of *Kitty Hawk*'s steam catapult sends ROMPER 502 slamming into the void. Moments later, the other 'cat' sends wingman ROMPER 507 aloft. The two-plane strike force heads on a predetermined course away from the ship, guided by instructions from an E-1B Tracer. The TARCAP, or escort force of F-4 Phantoms, launches from *Kitty Hawk* after the Intruders but soon passes them heading towards the beach.

43

It matters not whether it is sand, dirt or granite (although in North Vietnam it is most likely to be powdery dirt pushing up against scrub); anywhere in the world naval aviators call it crossing the beach. As the two Intruders pass the coast at 2235 hours, they descend and separate to follow different headings toward the target, each down on the deck, in the weeds, hoping to pose the smallest possible target to the enemy's field-sweep radar. Aboard ROMPER 502, the BN soon receives radar warning indications. The enemy's defensive net is attempting to pinpoint their ingress, but the Intruder is already well on its way to the target.

Rain spatters the windshield in the nocturnal dark. For reasons known only to themselves, the designers of the Intruder have provided the pilot with a windshield wiper, but not the BN. (The 'wiper' is actually a blown-air system, not a blade.) The prospect of a thunderstorm gouging up its torrents of water and electrical eruptions remains an 'unknown'. This is very much Intruder weather.

North Vietnamese MiGs come out to play. They are immediately engaged by the TARCAP Phantoms. It does not pay to disagree with a Phantom. No enemy fighters get within striking range of the incoming Intruders.

When not protected by Phantoms or by its natural cloak, this being the poor weather other aircraft fly poorly in, the A-6A Intruder *is* vulnerable. If the A-6A were caught in a dogfight with a MiG-17 or MiG-21, it would have a tough time and would have to jettison its bombs to survive. Intruder crews are not expected to practice fighter tactics in the way that fighter pilots do. The remarkable array of black boxes available to the pilot and BN is optimized for the strike role and offers only marginal help in detecting, anticipating and outwitting an enemy. The fact remains that despite losses, no Intruder has yet been downed by an enemy fighter.

But where the Intruder really shines is in its internal suite of equipment for the air-to-ground role against targets like the North Vietnamese bridge being attacked today. The bulbous shape of the Intruder's nose was dictated by the radars and NWDS (navigation and weapons delivery system) needed to carry out the night and all-weather attack mission. Wedged in their A-6A Intruder maze of gadgetry, not unlike a cockpit in a science-fiction space ship, Claney and Parker bore relentlessly toward the target at perilous altitude—most of their decisions being made for them by black boxes.

If a man can be comfortable heading into harm's

way, our Intruder crew is comfortable. Their cabin even has a special storage facility for box lunches and four 1-quart thermos bottles! So far tonight, the enemy is too slow and too surprised to track the men and fire effectively with SAMs and AAA, but it *can* happen. On its second cruise, VA-85 lost aircraft 151590 over North Vietnam on 19 January 1967. 151587 and 152589 were also lost by the squadron over North Vietnam, on 24 March 1967 and 24 April 1967 respectively. The losses are no joke. Parker remembers a raid over North Vietnam when an Intruder took the enemy's wrath. 'We were getting *Fire Can* indications [warnings of being stalked by the radar associated with 85-mm AAA guns]. All of a sudden, there was flak everywhere. This one orange burst ripped the wing off an Intruder and sent it spinning end over end, like an acorn tumbling from a tree. Fire swept over the Intruder. The crew couldn't get out. Those guys were incinerated . . .' Tonight, finally within moments of their target, Claney and Parker intend to avoid a similar fate.

On target . . . Boring in at 400 mph (640 km/h), ROMPER 502 hugs the enemy's terrain, its black boxes making subtle adjustments for each gully and hillock which could transform the Intruder into a trashed maze of broken, burning steel. Tim Parker

has picked up the bridge ahead on his computerized attack system. Ike Claney brings the A-6A to a computer-selected IP (initial point) some ten miles (16 km) from the target. Claney is using an instrument panel with a mix of old and new, traditional dials surrounding his vertical display indicator and enhanced optical sight. Parker: 'At this point, there's a lot of chatter between the two of us. We're hyped up, and looking for threats.' Whisps of burning red dance around the Intruder as small-calibre guns open up. As often happens, Ho Chi Minh's defences are too few, too late. 'We got it', Claney says on ICS. Parker begins ticking off the distances from a grid on his readout. 'Seven miles . . . ten degrees left, Ike . . . Six . . .' Intruder and bridge are coming together.

It's the pilot who drops the bombs and he has two ways to do it. On the rare instances when he makes a manual drop, he pushes the 'pickle button', a teat

The war placed heavy demands on the Replenishment Air Groups (RAGs) which trained Intruder crews, including the 'Green Pawns' of VA-42 stationed at NAS Oceana, Virginia. 'Pawns' A-6A number 155626, coded AD-532, pauses during a visit to Scott AFB, Illinois, on 22 August 1970
(Fred Roos)

located on top of the control stick. Usually, however—and today—using the 'system', the pilot pulls a trigger on the stick known as a 'commit switch', in effect authorizing the computer to release the bombs at exactly the correct instant.

There is no dramatic climax. The pilot simply commits. The computer makes the drop. No one needs to say, 'Bombs away' and no one does. Both men feel it instantly. The A-6A gives a slight but noticeable tremble as the 500-pounders go off in a string at the bridge. The seemingly unclimactic bomb drop is followed by sudden, violent manoeuvre as Ike Claney is now free to 'do that pilot shit', as the BN would call it—jinking and zigging to avoid the criss-crossing gunfire around the bridge.

With the enemy's gunners riveting on our fast-moving A-6A, the second Intruder, ROMPER 507, arrived from a different direction and unloads its bombs. Then 507, too, takes evasive action while both pilots feel their way back toward treetop level.

Bombs flash against the structure of the bridge, twisting and tearing, dropping spans into the river.

Heading out, Claney receives welcome word from the Tracer that no MiGs are now in the area. The enemy's efforts to hit the Intruder with ground fire are, again, too little and too late. 'Heading for the beach', Claney intones on ICS. On egress, the crew exchanges information in short, choppy sentences and often goes for long periods without speaking at all.

Aircraft 149937 (left) was employed to test a flight refuelling system in 1966, but at the time the Navy was not enthusiastic. The prototype KA-6D Intruder tanker aircraft did not make its first flight until four years later (Grumman)

Today, scattered low-level clouds give way to the anticipated thunderheads. Rain becomes sleet. Lightning bursts around the Intruder. Claney and Patterson are going home in an element which is the special world of Intruder people. The Black Falcons have done their job. Against moderate defences—much less than they expected—their precision attack has felled the bridge, successfully completing *Kitty Hawk*'s mission. Tomorrow, perhaps, a morning mission will stymy enemy efforts at repair. Perhaps an RF-8A Crusader escorted by Phantoms will fly a post-recce mission and bring back photos to confirm the damage. At exactly midnight, Ike Claney guides ROMPER 502 across the beach and climbs to 5,000 ft (1524 m) now that enemy defences are behind. The E-1B Tracer confirms that they are coming out with no MiGs behind them.

Friday: 0100 hours

ROMPER 502 and ROMPER 507 join formation off the coast, take on added fuel from a KA-3B Skywar-

The Marines had to be prepared to operate their aircraft under primitive conditions, especially at Chu Lai which was all pierced-steel planking or SATS (Short Airfield Tactical Support). On 1 May 1968, A-6A aircraft 152587, coded CE-3, or VMA(AW)-225 operates from a SATS strip with a full ordnance load
(USMC)

rior tanker, and settle for their approach, making a straight-in final leg toward *Kitty Hawk*'s heaving deck. Studies have shown that men like Claney and Parker have some of their tensest moments during approach, especially at night in rough seas. A thousand yards (960 m) from the moving carrier, Claney talks to the LSO (landing signal officer) and is instructed to 'call the ball'. The ball is a series of lights mounted vertically to be lined up with a horizontal row of green datum lights, thus aligning a plane on a proper glide path. 'Ike has the ball', our pilot replies. Gear down, flaps down, hook down, the A-6A Intruder comes in for a solid, constant-attitude landing.

The huge weight of the Intruder catches the wire and the sturdy craft slams down on the carrier deck. Our crew has pulled it off. Claney follows hand-held torches in the hands of a yellow-shirted deck hand and parks the airplane. The mission is done—except for an hour or so of debriefing. This will end, as the preflight brief ended, with the two men alone comparing notes on how they'll do it next time.

Combat Veteran

If the Claney/Parker mission seems remarkable, it is perhaps because the Intruder's mission is, as Parker readily admits, 'rather like that old story of hours of boredom interrupted by brief moments of sheer terror'.

Intruder missions often involved enormous risk but received practically no attention. On the pitch-black night of 16 March 1967, a single A-6A Intruder with Commander R J Hays, CO of VA-85 as pilot and Lieutenant Ted Been as bombardier-navigator made a very successful night radar attack on the Bac Giang power plant, 23 miles (40 km) northwest of Hanoi. Despite heavy AAA fire and a number of SA-2 firings, Commander Hayes was able to deliver over 12,000 lbs of ordnance to his target. That same month, VA-85 flew a successful mission against the heavily defended Thai Nguyen steel works.

Needless to say, with a system like the A-6A Intruder, which represented a leap forward in electronics technology and which began its career with maintenance difficulties, the all-important technician aboard the carrier (or, for the Marines, on the ground at garden spots like Chu Lai) was absolutely critical. In 1965–68—that is, during the heavy fighting of the Rolling Thunder campaign, before President Johnson halted operations over North Vietnam with a stroke of the pen—the draft contributed to the US' good fortune in having a superb cadre of specialists who kept the Intruders flying.

TOP LEFT
LCDR Tom Patterson is typical of the bombadier-navigators who fly in the right-hand seat of the Intruder. Tom completed BN training towards the end of the Southeast Asia conflict and made his first cruise aboard USS Coral Sea *(CVA-43) just as US participation in the Vietnam fighting ended in January 1973*
(courtesy LCDR Patterson)

LEFT
A-6A Intruder carrying 'buddy' refuelling pack aboard USS Coral Sea *(CVA-43) just after the January 1973 cessation of the US combat role in South Vietnam*
(LCDR Tom Patterson)

ABOVE
Although assigned to the Atlantic Fleet, USS Saratoga *fought in Southeast Asia and eventually acquired the first Intruder fleet squadron, VA-75. Sara was also the first carrier to take on the added mission of anti-submarine warfare and to change from a CVA (attack carrier) to CV (all-purpose carrier). KA-6D number 151589, coded AC-521, wears post-war markings as it arrives from CV-60 for a landing at NAS Roosevelt Roads, Puerto Rico, on 20 July 1979*
(USN via R J Mills Jr)

Consider the 'Bats' of VMA(AW)-242, the little-publicized Marines who became the first in the Corps to fly the Intruder. VMA(AW)-242 took delivery of its first A-6A airframes at NAS Oceana, Virginia, in mid-1964. At the time, the squadron was commanded by Lieutenant Colonel Robert Wilson. Soon, it was providing training to the Marines' other Intruder squadrons, including VMA(AW)-533 which went to Chu Lai. Had it not been for an exceptional cadre of technicians, the squadron simply would not have been able to do its job. One of Wilson's successors as skipper of the Bats, Lt Col Earl E Jacobson, was merely uttering the obvious when he said, 'There would be no Rolling Thunder missions or any others for that matter if it weren't for the men on the ground with the tools in their hands and the knowledge in their heads . . .'

It was through no fault of Navy and Marine Intruder crews that the policy decision was made to temporarily end the bombing of North Vietnam. President Johnson's 31 October 1968 bombing halt, one of the results of his withdrawal from that year's presidential campaign, came as a blow to many of the aircrews who'd been flying north, following the words of a popular Petula Clark song urging them to go 'downtown' (to Hanoi) and taking enormous risk in the hope that their efforts would accomplish

something. The bombing halt, aimed at prodding Hanoi's leaders into negotiations, instead accomplished something, all right, but not what anybody wanted: it gave the enemy, reeling from the impact of three years of air strikes, time to reinforce and expand his air defence network. When it became necessary to go back to North Vietnam in 1972, this time with Richard Nixon in the White House, the crews of Intruders and other strike aircraft learned that the enemy had succeeded in vastly rebuilding his defences. In 1972, Navy and Marine Intruder crews faced North Vietnamese missiles, MiGs and Triple-A which were far better prepared and more formidable than ever before.

TOP LEFT
Fitted with long-range fuel tanks, an A-6E TRAM of VA-95 'Green Lizards' from the deck of USS Enterprise *(CVN-65) escorts a majestic Tupolev Tu-142* Bear D *reconnaissance bomber of Soviet Naval Aviation over the Sea of Japan on 10 January 1982*
(USN)

BOTTOM LEFT
VA-95 rack up another successful intercept as bureau number 161230 keeps an eye on Tupolev Tu-16 Badger, *serial 1883207, tail number 24. The A-6F will have provision for Sidewinder AAMs fitted as standard, providing a welcome self-defence capability for the aircraft and additional air defence protection for the carrier battle group. You don't have to be 'Duke' Cunningham to splash a large subsonic bomber*
(USN)

Chapter 3
EA—6B Prowler

The EA-6B Prowler is in every important respect an entirely different aircraft type from the A-6 Intruder. The Prowler almost certainly deserves a volume of its own, being one of the principal electronic warfare (EW) aircraft in the world today. The Prowler's primary mission is to protect friendly surface vessels and aircraft by jamming enemy radars and communications. Indeed, were it not for the Air Force's EF-111A Raven, which performs a somewhat similar mission and was itself modified by Grumman using EA-6B Prowler experience, there would exist in this world *no* real competition for the Prowler at all. The EA-6B is, in the true meaning of the word, unique.

Comparison with the EF-111A Raven is not exactly apt, since the latter system came along many years later, but it could be pointed out that both machines have a bulged fin fairing enclosing sensitive surveillance receivers that can detect enemy radars at long range. The AN/ALQ-99 tactical jamming system is the principal electronic weapon carried by both Raven and Prowler. On the Prowler, the capability exists to carry five integrally powered pods, driven by what look like tiny propellers in front, with a total of ten jamming transmitters. Each pod is dedicated to one of seven frequency bands.

A stretched, four-seat development of the Intruder, the bigger Prowler employs the more powerful 11,200-lb (5080-kg) static thrust Pratt & Whitney J52-P-408 engine. The powerplant change was made from the 22nd airframe onwards, the first twenty-one EA-6B Prowlers having been powered by twin J52-P-8A engines of 9,300 lb (4218 kg) static thrust. The Prowler is perhaps not as large as it appears when viewed for the first time, but a 40-in (103-cm) extension of the basic Intruder fuselage makes room for two more crew members and, more importantly, a host of electronics gear.

Dimensions for the EA-6B Prowler differ from those of the A-6E Intruder as follows: fuselage length is 59 ft 5 in (18.11 m). Height is 16 ft 3 in (4.95 m).

The wheelbase is 17 ft 3 in (5.26 m) and width with wings folded is 25 ft 10 in (7.87 m). In addition to the pilot, the Prowler carries three naval flight officers (NFOs) known aboard this aircraft type as 'Ekmos', slang for electronic countermeasures officer (ECMO). These hardy souls, valiant enough to head into battle essentially as passengers on a fast-jet aircraft likely to be a major target to any enemy, man a complex array of jamming, deception and other ECM equipment designed to confuse and disrupt an enemy's defences.

The pilot is responsible for an aircraft which handles very differently from the Intruder. In early days, a cross was painted on the front of the EA-6B's radome so that landing signal officers (LSOs) would not confuse the longer, heavier airplane with the Intruder. That practice is still followed in some carrier air wings, in the Atlantic Fleet at least. The EA-6B Prowler pilot may not spend as much time hugging the Earth, but in most respects his job is at least as challenging as any Intruder jockey's.

Unlike many other electronic warfare platforms including the F-4G Phantom intended for close-in attacks on enemy radar sites, the EA-6B carries no offensive ordnance and is thus limited to its electronic role as a 'weapon'.

Electronic Warrior

The EA-6B Prowler was developed from the two-seat EA-6A variant of the Intruder, which is treated elsewhere in this text. The EA-6A did not seem to offer

Not even Grumman's own History Center had photos of the little-known EA-6B Prowler full-scale mockup until researcher Jim Wogstad helped the manufacturer's experts to unearth them. The mockup was remarkably true to the real EA-6B configuration which followed (Grumman)

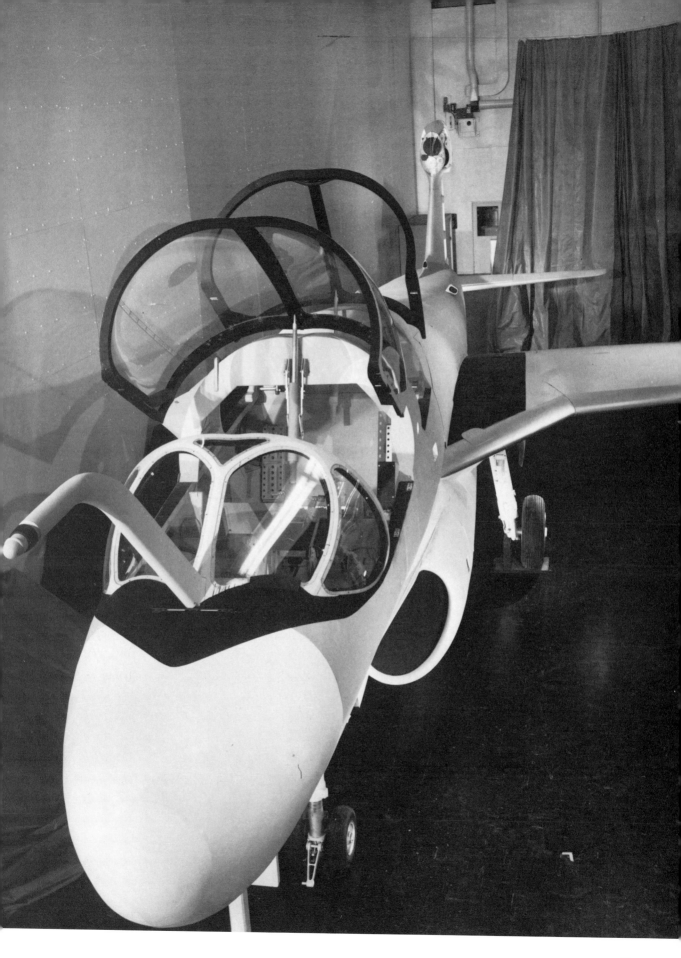

the opportunities for growth and further development that might come with a larger, sturdier machine. Grumman thus submitted a concept design proposal (CDP) in June 1964 for a larger aircraft capable of detecting and jamming known Soviet radars. Larry Mead's design team in Bethpage looked, once again, at a variety of possibilities, including a three-seat version of the Intruder. In the end, a four-seater, with the basic Intruder fuselage lengthened by 40 inches (103 cm) was decided upon. Grumman completed a full-scale mockup constructed primarily of wood, along with a variety of wind-tunnel models which, when evaluated, proved the basic soundness of the design.

Known to the company as Model 128J, the first EA-6B Prowler was completed from an A-6A airframe (bureau number 149481, originally A-6A number 15) and was flown at Calverton by Don King on 25 May 1968. This airframe is sometimes referred to as article M-1 in Grumman documents. It was soon followed by test article M-2 (bureau number 149479, originally A-6A number 13) which was used for weapons systems flight evaluation. The US fiscal year 1969 military budget allocated $139 million—a real bargain, even in those pre-inflation days—for the initial purchase of eight full-scale development and production EA-6B aircraft and systems.

Board of Inspection and Survey (BIS) trials of the Prowler were carried out in 1970 and, that same year, a developmental airplane performed well in carrier qualification (carqual) tests aboard the USS *Midway* (CVA-41). There followed a lengthy working-up period before operational deployment of the Navy's first two Prowler squadrons in 1972. These squadrons ended up carrying out 720 combat sorties in support of Linebacker missions over North Vietnam. Deployment of the EA-6B Prowler to other units in the Fleet was speeded-up when it was recognized that the aircraft offered unique capabilities. At first, recruiting of 'Ekmos' was difficult, but the Prowler's mission quickly gained a reputation as being a challenging one to fly. Since the late 1960s, EA-6B Prowlers have rolled off the Grumman production line at the seemingly leisurely rate, actually itself the result of much planning, of six airframes per month.

Prowler Variants

Within the EA-6B Prowler community, no fewer than five sub-varieties of the basic airplane exist or are comtemplated. The 'basic' EA-6B Prowler first went into combat against North Vietnam in July 1972 with the 'Lancers' of VAQ-131 aboard USS *Enterprise* (CVAN-65), the men of this squadron having

EA-6A, coded GD-14 from VAQ-33 'Firebirds', stands ready for her crew at NAS Oceana on 30 April 1983 (Don Linn)

been trained at NAS Whidbey Island, Washington by the Prowler community's replenishment air group (RAG), VAQ-129. The 'Scorpions' of VAQ-132 and the 'Garudas' of VAQ-134 quickly followed to the Gulf of Tonkin where Prowlers flew some the aforementioned 720 combat sorties, remaining in the battle zone through the 'Eleven Day War' of 18–29 December 1972—the 'Christmas bombing' to its detractors, Linebacker II to its advocates—which finally brought North Vietnam to the conference table.

Although they apparently operated within range of North Vietnam's formidable defences, *not a single EA-6B Prowler was lost in combat or to operational causes in the Southeast Asia conflict.*

The EA-6B continued to be active in immediate post-war cruises by carriers like USS *Coral Sea* (CVA-43) and in the April 1975 evacuation of Saigon and the *Mayaguez* incident which followed.

The second generation, the EA-6B EXCAP (extended capability) Prowlers differ from the first generation or 'basic' Prowler in being able to carry out the electronic warfare role against enemy radars in eight frequency bands instead of four. Furthermore, EXCAP Prowlers have an increased number of computerized systems. After a further four airframes brought to six the number employed in the Prowler test programme, and following delivery of 23 'basic' Prowlers which were known at Grumman as articles MP-6 through MP-28 (bureau numbers 158029/040; 158540/547; 158649/651), the first aircraft in the EXCAP series was delivered.

Known as Grumman's article P-29, bureau number 158799 reached the Fleet in January 1973, the month of the end of the Vietnam war, and was the first of twenty-five Prowlers in the EXCAP series. The EXCAP version first deployed to the Mediterranean in January 1974 and by January 1976 was operating with the Pacific Fleet as well.

EA-6B ICAP (improved capability) Prowlers came next, ICAP 1 being the Navy's acroynm for the third of five Prowler variants. Beginning with Grumman production article P-54 (bureau number 159907), which made its first flight in July 1975 forty-five ICAP 1 airplanes were delivered, the first test article

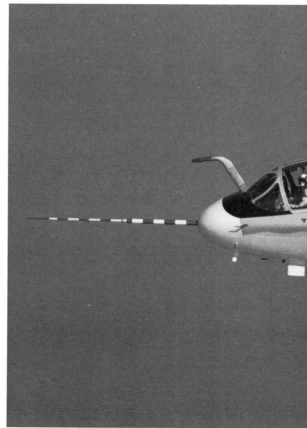

TOP RIGHT
With Don King at the controls, Grumman article M-1, the very first EA-6B Prowler (bureau number 149481) makes a shallow turn on its first flight on 25 May 1968 at the manufacturer's Calverton, New York facility
(Grumman)

RIGHT
First flight of the EA-6B Prowler. Don King eyes the chase plane as he puts article M-1, alias bureau number 149481, through its paces. The 'candy stripe' nose pitot tube was deleted on production airplanes
(Grumman)

TOP

586 dressed for Indy. *EA-6B Prowler EXCAP aircraft,
bureau number 159586, is seen in the markings of the 'Yellow
Jackets' of VAQ-138 when the squadron was aboard the
USS* Independence *(CVA-62) in the mid-seventies.
Wearing side number AJ-616, the Prowler is depicted at
NAS North Island, California on 8 August 1976*
(USN/Bruce R Trombecky)

ABOVE

*586 dressed for Ike. Same aircraft, same squadron, but now
side number AG-601 appears on EA-6B Prowler 159586 of
the 'Yellow Jackets' as it prepares to go to sea aboard USS*
Dwight D Eisenhower *(CVN-69). Now the Prowler is seen
at NAS Oceana, Virginia, on 11 May 1978*
(USN/J E Michaels)

appearing in February 1976 with digitally tuned receivers, new crew instrument displays, and redesigned cockpit configurations.

A distinct recognition feature for aircraft beginning with the ICAP 1 machines and continuing onward is the 'sawtooth' found in the Prowler's L-shaped nose air-refuelling probe, this being caused by locating an AN/ALQ-125 radar receiver in the base of the probe. Initial delivery of the full-fledged production ICAP 1 machines began in September 1977.

EA-6B ICAP 2 variants—these are fourth in the Prowler family parade—appeared beginning in 1983 with the delivery of Grumman production article P-99, or bureau number 1611776. Again, the saw-toothed flight-refuelling probe appears. In the ICAP 2 variant, the earlier AYA-6 computer was replaced with the more advanced AN/AYK-14. Added, too, was a capability to link two Prowlers via

Tacan datalink to enable two airframes to work together in a coordinated EW thrust against an enemy's defences. These features of the ICAP 2 variant will be discussed later when a sample history of a typical EA-6B Prowler squadron over a one-year period is presented. Suffice to say that improvements make the Prowler far more effective, enhance work conditions for the somewhat crowded crew, and enable the EA-6B to perform a wider range of electronic warfare missions.

Finally, the fifth member of the family, and the one which faces current budget difficulties in a deficit-conscious Congress is the EA-6B ADVCAP (advanced capability) Prowler aircraft which has long been scheduled for the late 1980s—although the status of funding for ADVCAP was very much undecided as this volume was in preparation. Some critics were contending that, while the improvements are significant, the cost—not yet finalized—is out of pro-

The essence of EA-6B. Its side number hastily changed from the 600 series to the first in the 620 series during the rapid shuffle of the period, this EA-6B ICAP2 Prowler of the 'Playboys' of VMAQ-2 is towed into position aboard USS Coral Sea (CV-43) on 29 January 1986. Two months later, this Prowler participated in combat operations against terrorist-related targets in Libya (USN/PH2 Rory Knepp)

portion. ADVCAP was launched in fiscal year 1983 with a development contract awarded to Grumman. Under the terms of the contract, a substantial part of the work was to be carried out by Litton-Amecon, which was developing a new receiver/processor system based on the AN/ALR-73 system long employed by the E-2C Hawkeye. Flight of the first ADVCAP airframe has been delayed, apparently by continuing budgetary debate and actual production is not expected before 1990. Total cost of the ADVCAP programme is being estimated at no less than $400 million. It is probably a bargain.

The ADVCAP variant will add jamming and passive detection gear to the basic Prowler electronics suite. The basic four-man Prowler configuration will remain unchanged. It is thought that, possibly, consideration may also be given to arming the Prowler for the first time, perhaps with AGM-88A HARM (high-speed anti-radiation) missiles. Advanced Pro-

wlers in production in the 1990s may also find themselves, for the first time, equipped to carry the AIM-9L Sidewinder or AIM-120A AMRAAM (advanced medium-range air-to-air missile).

Navy men put a high premium on the importance of funding for the ADVCAP Prowler, which is likely to face resistance in a budget-conscious Congress. Though not even its most ardent supporters claim it will be cheap, many argue that it is the best way the Navy can exploit the relationship between the electronic warfare threat (that is, the bad guys' technology) and the ability to counter that threat (the good guys). While the existing receiver processor on the ICAP II Prowler aircraft is essentially unchanged since 1971, the ADVCAP Prowler's Receiver Processor Group (RPG) will be much more discriminate, much more precise, and able to look at all of the frequencies being employed by an enemy: it will be fast enough to find the critical signal at any frequency. The exact details of improvements in the ADVCAP Prowler's system are both too arcane and too sensitive for discussion in any real detail. Suffice to say that the ADVCAP airplane will be a quantum leap over its predecessors. An essay by Commander Michael D Barrett in a staunchly pro-Navy publication terms the ADVCAP Prowler absolutely essential.

It would be accurate to make the point—and Prowler men do it with some pride—that the EA-6B Prowler has had an active role in virtually every cold-war

conflict since the end of the Vietnam fighting. Else-where in this volume, the role of the Intruder in 1983 operations in Grenada and Lebanon will receive some attention. Suffice to note that the EA-6B was in the sky while those actions were unfolding, including the controversy-ridden 4 December 1983 Alpha strike in Lebanon which was far from a success. In a more warmly-remembered example, and one which will perforce be mentioned again later, the EA-6B fouled communications during the 1985 night intercept by F-14A Tomcats of the Egyptian airliner carrying the hijackers of the Italian cruise vessel *Achille Lauro* and forced the Boeing 737 to land at NAS Sigonella, Italy. A hassle with Italian guards and the plain fact of Italian sovereignty prevented the carefully-envisioned second step of the *Achille Lauro* affair—hustling the captured hijackers aboard the Delta Force C-141 Starlifter that landed at Sigonella *behind* them and spiriting them rapidly to McGuire AFB, New Jersey—but at least the bad guys made it as far as a courtroom in Rome. Diplomacy may have hamstrung the efforts of warriors as happens so pain-fully often but, in any event, the EA-6B had unques-tionably done a superb job. As an encore to this performance, Prowler squadrons aboard three car-riers in the Mediterranean performed nobly in the March and April 1986 actions against Libya, of which more will be heard later in this narrative. (See page 159).

Previously equipped with EAK-3B Skywarriors, the 'Scorpions' of VAQ-132 moved to NAS Whidbey Island, Washington, on 5 January 1971 and transitioned into the EA-6B Prowler. On 12 July 1972, in the Tonkin Gulf, the squadron flew the EA-6B in combat for the first time. In the mid-1980s, under Commander R S Weber, the squadron moved from USS Forrestal *(CV-59) to USS* Eisenhower *(CVN-69). EA-6B aircraft 161244, modex AE, side number 605, is shown during the* Forrestal *period*
(USN)

RIGHT
Used with permission from the magazine, the insignia of the 'Playboys' of VMAQ-2 adorns the antenna-studded tail of EA-6B Prowler 159909, side number AB-605. The Prowler was board the USS America *(CV-66) during combat operations against Libya in April 1986*
(Dave Parsons)

Prowler Squadrons

The EA-6B Prowler has been assigned to no fewer than an even dozen US Navy squadrons (VAQ-129 through VAQ-140, with a thirteenth, VAQ-141, about to be added to the expanding Navy), and one Marine Corps squadron (VMAQ-2) which also oper-ates the aircraft on carrier decks.

With the 'Playboy' bunny on their airplanes (permission of Hugh Hefner), the Cherry Point, North Carolina-based Marines of VMAQ-2 flew the two-seat EA-6A Intruder for many years before graduating to its four-seat relative. Although nominally an east coast unit, VMAQ-2 began training in the EA-6B in September 1977 at NAS Whidbey Island, Washington, where all Prowler work-ups take place and where VAQ-129 serves as the Prowler Fleet Replenishment Squadron (FRS), what aircrews call the RAG.

Already having enjoyed a remarkably good record with the two-seat EA-6A Electric Intruder, the Marines seemed to 'take' to the Prowler with uncommon ease. VMAQ-2 made an especially good showing during operations aboard the nuclear-powered USS *Nimitz* (CVN-68) during the September–October 1981 NATO exercise known as Display Determination 81. VMAQ-2's Marines, who have resisted all efforts by an entrenched bureaucracy to change their Playboy rabbit nickname and insignia, are especially proud of the contribution they make on board Navy carriers—one which was especially laudable during 1986 operations against Libya. Normal complement of a squadron's detachment on board a fleet is no more than four airplanes and at times even fewer.

Crew Training

Pilot of an EA-6B Prowler, like his colleague in the Intruder, begins his career with a college degree and a commission in the Navy. Some, of course, come from the ranks of the US Navy Academy at Annopolis. At present, he learns how to fly in the Beech T-34C and then progresses to two jet aircraft, first the North American T-2C and next the Douglas TA-4J Skyhawk. By the early 1990s, both of these types will have been replaced by the T-45 Goshawk, the Douglas-built naval adaptation of Britain's highly successful Hawk trainer.

Subject of course to the Navy's needs, near the end of the TA-4J syllabus, 'guys with the best grades get the first choice', as Captain Ken (Doc) Koskella puts it. *Affecionados* of the EA-6B Prowler need not have their feelings hurt if it is noted that most 'first choices' usually end up being another Grumman product— the F-14 Tomcat—but not always. Koskella, the already-introduced flight surgeon who is head of the F-14D-A-6F Program at NATC Patuxent River, Maryland, says that 'there is a peculiar breed of animal who always wanted to fly in this kind of big fast jet and who is remarkably good at it'.

The typical naval flight officer who becomes one of the three 'Ekmos' aboard an EA-6B Prowler is expected to start with a college degree and to attend the basic flight school at NAS Pensacola, Florida, which is also the starting point for Intruder bombardier-navigators. The course at Pensacola, as described to the author by Lieutenant Jim Gilgliotti, takes ten months and is a course in basic airmanship, basic skills, and radar navigation. At a major juncture in the training syllabus, the men who will serve as

Low-visibility gray in the snow. Joe Handelman's finely-attuned lens captured EA-6B Prowler 161351, coded NK-604, from the decks of USS Constellation (CV-64). The Prowler was parked on a snow-covered ramp at Andrews AFB, Maryland, on 21 January 1984
(Joseph G Handelman, DDS)

A rare view of two EA-6B Prowlers in close formation carrying a mix of jamming pods and fuel tanks. VAQ-138 'Yellow Jackets' were embarked on the USS Nimitz (CVN-68) when this photo was taken on 10 November 1986
(USN/PH1 Ronald Beno)

A rather smart EA-6B from VAQ-134 'Garudas' receives attention from maintenance men on the USS Enterprise *(CVN-65) during Exercise Rimpac in the Pacific Ocean on 13 April 1978*
(USN/PH2 P J Salesi)

ABOVE
The crew of a Prowler from VAQ-138 find time to help compose this pleasing study for photographer PH1 Ronald Beno during a cruise on Nimitz *in 1986. Bureau number is 161118*
(USN/PH1 Ronald Beno)

crew members aboard 'slow mover' P-3C Orions separate from the remainder of the group, who will fly fast jets. The prospective 'Ekmos' then go on to do terrain and radar work.

Once chosen for the EA-6B Prowler, the electronic warfare officer proceeds to the replenishment air group, the 'Vikings' of VAQ-129 at NAS Whidbey Island, Washington. Once mated to the actual aircraft they'll fly, the men learn how the crew of a four-seat Prowler can function as a team. The training syllabus at the RAG is also likely to include a considerable amount of flying around in the back seat of a TC-4C, the Navy's version of the Grumman Gulfstream G-I executive aircraft (officially assigned the nickname Academe, which no one ever uses), configured with an Intruder nose and furnished with rear-fuselage stations for the BN/ECMO trainee. The TC-4C is one of those quiet success stories which will warrant a few words in its own behalf later in this narrative.

Workups in the TC-4C and in the EA-6B Prowler itself are followed by initial assignment to a Fleet squadron for a period of two to three years. On assignment with the squadron, the newly arrived 'nugget' ECMO is placed under the charge of a more experienced Prowler crew member who helps him to get settled into carrier life and Prowler operations. The ECMO community is a very small cadre of dedicated people and, over time, nearly all of them get to know each other.

Prowler Workload

In fact, the EA-6B workload is split among three crewmen and although each man learns all prospective tasks, the work performed depends on where you're sitting in the aircraft. In the 'basic' and EXCAP Prowlers, as an illustration, 'Ekmo 1' in the right seat beside the pilot operates half of the tactical jamming system (TJS) while 'Ekmo 2', directly behind him on the starboard side of the airplane, covers the other half, thus assuring efficient frequency coverage within operational environments that feature a high density of signals, 'Ekmo 3', on the port side behind the pilot, operates the AN/ALQ-92 communications jamming system. The men communicate on an ICS (intercom system) similar to that found aboard the two-seat Intruder, and when an active mission is being flown, there is considerable discussion and interplay.

EA-6B on the Boat

It is not at all easy for an EA-6B Prowler unit to put to sea, in part because of the enormous amount of support and supply material which must accompany this electronic wonder. Three equipment vans must accompany a VAQ squadron on a shipboard cruise, carrying highly specialized electronics monitoring and test equipment which is essential to keeping a Prowler force in the air. The vans, which measure about 8 × 40 feet, are usually suspended from the hangar deck overhead and are often viewed as an annoyance by members of other squadrons who resent the extra space taken up by the Prowler people.

Like all US Navy aircraft of the period, the EA-6B Prowler began its career in the colourful markings that once typified carrier-based squadrons. At the international air tattoo at RAF Greenham Common, England in June 1979, for example, EXCAP Prowler 159584, side number AG-602, belonging to the 'Yellow Jackets' of VAQ-138 aboard USS Dwight D Eisenhower (CVN-68) was resplendent in the colours of carrier air wing seven (CVW-7) with the familiar yellowjacket painted on its rudder. For a time, squadrons seemed to be in fierce competition to produce the most colourful Prowlers, to the delight of camera bugs. By the early 1980s, however, naval aviation planners had been sold on the idea of 'toning down' their airplanes in low-visibility paint schemes. Sadly for us all, except those of us who care about our guys being spotted and sandbagged by the baddies, gray became good and the EA-6B Prowler quickly fell in with this trend.

Force 'Multiplier'

Navy people in the EA-6B community consider their aircraft to be a 'force multiplier' which enhances the effectiveness of a carrier battle group and of the entire Fleet.

The EA-6B's principal mission is to provide a naval task force commander with a kind of double shield—a defensive shield to hide the task force and to jam the sensors of long-range bombers or cruise missiles launched against the fleet, and an offensive shield to protect the carrier's strike aircraft from the air-defence systems of enemy surface ships or defended points ashore (North Vietnamese anti-aircraft guns and Libyan SA-3 missile sites being two historical examples of the latter).

The EA-6B Prowler's TJS (tactical jamming system) is an AN/ALQ-99E; the Prowler can carry five integrally-powered pods with a total of ten very high-power 'smart' jamming transmitters. Each pod carries equipment to cover one of seven frequency bands. The Prowler can carry any mix of pods or fuel tanks, depending on the mission. The surveillance reviewer located in the EA-6B's distinctive tail-fin pod picks up threat radar emissions at long range and feeds this threat data into the central digital computer for processing. Detection, identification, direction-finding and jammer set-on sequence are performed automatically or by crew assist.

Antenna Locations

The EA-6B has six antennas located in its tail fairing for the ALQ-99. Fairings on the fin house L-F band ECM receiving antennas. The antenna for the ALQ-100 deception ECM system protrudes from the left

TOP
*War, Electric Intruder (left) and attack Intruder (right).
At NAS Atsugi, Japan circa. 1968, at the height of
Southeast Asia fighting, EA-6A Electric Intruder 151591,
coded RM-12, of squadron VMCJ-1 poses with A-6A 'mud-
moving' Intruder 154160, coded ED-3, of squadron
VMA(AW)-533. Flying from Chu Lai, 154160 was lost in
a combat mission over Laos on 17 March 1969*
(Nicholas M Williams)

ABOVE
*A-6A Intruder in combat. Although a number of A-6As
aboard USS Constellation (CVA-64) wore an
experimental camouflage scheme during combat operations in
the Gulf of Tonkin in 1965, this particular paint scheme—
with blue undersides—is extremely rare. Seen on 7 September
1965, aircraft 149948 appears to be testing the colour pattern*
(USN)

Launching from USS Ranger *(CVA-61), which brought Intruders to 'hot spots' in Vietnam and Korea, A-6A number 152906, coded NE-506, belongs to the 'Boomers' of VA-165 which employed the aircraft early in combat. Early Intruder radomes were black*
(USN)

At the Strike Aircraft Directorate ('Strike', for short), NATC Patuxent River, Maryland on 27 February 1987, LT Dave (Cleve) Brown of Strike's F-14D and A-6F Program poses with friend. A-6E TRAM Intruder 157014, side number SD-503, is being employed in developmental work at Pax. Dave is an experienced bombardier-navigator
(Robert F Dorr)

ABOVE
*A-6E of the 'Golden Intruders' of VA-128, the west coast
Fleet Readiness Squadron, on a visit to Andrews AFB,
Maryland on 10 February 1979*
(Robert F Dorr)

RIGHT
*Heading to sea, KA-6D Intruder tanker 152910 of VA-165
is seen on carrier deck in Yokosuka, Japan, in early 1980*
(Tsunehiro Kouda)

TOP
*In the resplendent colours of the Naval Air Test Center,
EA-6B Prowler is seen at Strike Test, NAS Patuxent River,
Maryland on 31 March 1977*
(Robert F Dorr)

ABOVE
*The current low-visibility paint scheme of naval aircraft is
illustrated, with snow in background, by EA-6B Prowler
161351 coded NK-604 from USS Constellation (CV-64)
at Andrews AFB, Maryland on 21 January 1984*
(Joseph G Handelman, DDS)

outboard pylon. EA-6B ECM equipment includes ALQ-41 and ALQ-100 multiple-band track-breaking deception ECM (DECM) systems to spoof hostile radars. Additional antennas are located on the leading edge of the outboard pylons. The ALQ-92 communications jammer radiates through the large blade antenna below the chin of the aircraft. ALQ-41, ALQ-100 and ALQ-92 are manufactured by Sanders Associates.

The EA-6B's nose-mounted AN/APQ-129 search radar is roll- and pitch-stabilized to maintain the quality of the presentation during climbs, banks, and descents.

Notes compiled by author David A Anderton and quoted here with permission indicate that two EA-6B Prowlers flying a racetrack pattern at 25,000 ft (40,233 m) can meet a threat to the carrier from three directions. Using broadband brute-force jamming or modulated jamming, the Prowler pair can virtually deny the commander of an attacking Russian force any ability to detect the carrier group. In the absence of the EA-6B, according to simulation studies, a carrier group's air defence system can kill three attacking Tupolev Tu-22M *Backfire* bombers. This becomes six *Backfires* if one EA-6B is present, and eight *Backfires* if two EA-6Bs are present.

If attacking *Backfires* launch 36 missiles (in the simulation scenarios above), 31 will get through the screen if there are no EA-6Bs present and *all* will be locked on. With one EA-6B, 26 missiles will get through and 15 will be locked on. With two EA-6Bs, 26 missiles get through and only four are locked on. Other missiles from our Russian attack force are confused and fail to penetrate.

Anderton's look at simulation studies revealed that the EA-6B gives a carrier group's deck-launched interceptors (DLI) a few more minutes to arrive and fire AIM-54C Phoenix missiles at attacking *Backfires*. If each F-14 launches all of its missiles (five F-14s with six AIM-54C each, totalling thirty missiles), the total assumed *Backfire* force of one regiment would be wiped out.

The EA-6B works, when appropriate, with two E-2C Hawkeyes doing the passive tracking of targets, cooperative tracking and triangulation, developing estimates of the enemy's speed and heading. This teaming-up of EA-6B and E-2C can feed data to the F-14, who then computes a missile launch as if the F-14's own radar (assumed in this scenario to be jammed by the *Backfire* attackers) were working.

Vital Mission

In the actual conflicts where the Prowler has found itself committed—indeed, in all conflicts since World War 2—the US Navy has been fortunate enough to wage war without its own vessels coming under attack. Needless to say, however, exercises which involve attacks on American ships are far from academic. We live in a world of nuclear ordnance,

stand-off weapons, cruise missiles, and a variety of sea-skimming missile systems the best known of which is the French-developed Exocet. Since the 4 May 1982 action occurring in broad daylight during the Falklands war when no larger adversary than a lone Argentine Exocet missile made very short work of the Type 42 destroyer HMS *Sheffield* (D-80), no one has taken this threat lightly and the value of an electronic warfare platform like the EA-6B Prowler has been obvious. The 16 May 1987 Exocet attack by Iraqi Mirage F.1 strike fighters against the American guided missile cruiser USS *Stark* (FFG-31)—an incident reported as unintentional at the time, yet one which again raised questions about defending the Fleet from air attack—once more put the spotlight on the need to protect US Navy vessels from any attacking air force from as far out as possible.

Prowler Performance

Neither in appearance nor intent is the EA-6B Prowler any speedster. It is interesting to note, however, that in 'clean' configuration its 'never exceed' speed is as high as 817 mph (1315 km/h), suggesting that the portly Prowler might get somewhere near those numbers, though, certainly, it would have to happen in a very determined dive!

Ordinary maximum speed at sea level is 651 mph (1048 km/h) and the EA-6B cruises optimally at 481 mph (774 km/h). Maximum rate of climb is 12,900 feet (3932 m) per minute. Service ceiling is listed as 44,500 ft (13,565 m), although the author is certain Prowlers have worked at greater height. Combat range with maximum external fuel, says the brochure, is 2,399 miles (3861 km). That in-flight refuelling probe on the Prowler is all-important: its Intruder stablemate is the only dedicated tanker in the fleet and, at times, the Navy's tanker resources seem thinly spread, but the Prowler can and will be expected to cover vast distances on some of its electronic missions.

The Wizards

In a work of this length it is, inevitably, not possible to provide a detailed history of every squadron which flies an aircraft (sorry, guys!). And while every US Navy squadron is now tasked to prepare a history, other challenges must, in the course of events, take priorities. As a representative history on behalf of the entire Prowler community, one of the very best chronologies the author has seen was prepared by the 'Wizards' of VAQ-133, which in 1985 even included such details as the bureau numbers of the airframes assigned, these being EA-6B ICAP 2 Prowlers 161776, 161778, 161883 and 161884.

Much of naval aviation is the hard, gritty everyday work of operating ashore or deploying on a carrier. The Wizards' experience is typical. Commander William B Birkmaier Jr commanded the squadron

from 4 October 1983 to 26 April 1985, when he turned over the helm to Cdr Richard W Bennett, with Cdr Warren H Ide Jr taking the second slot as XO, or executive officer. All of these men have long experience and reflect the high level of talent found in the Prowler community. Although their aircraft crew roles are not recorded, the skipper and exec need not necessarily be pilots—it being worth repeating that the US Navy is far ahead of other services in providing command opportunities to the Naval Flight Officers who, aboard the Prowler, perform the 'Ekmo' role. It is no secret that weapons systems officers in the US Air Force's F-4, F-111 and F-15E communities do not routinely enjoy the same chances to take charge.

A year in the life of the Wizards, while providing little in the way of Hollywood drama, gives a picture of life with the Prowler including—that essential part of aviation—long absences from home and family. Here, then, on behalf of all Prowler squadrons, is the Wizards' story for 1985:

1 January–5 May: Naval Air Station, Whidbey Island, Washington. During this period, having just returned from a Western Pacific (Westpac) cruise on the nuclear-powered USS *Enterprise* (CVN-65), VAQ-133 finished converting from EXCAP to ICAP 2 airframes and began training for a future Westpac deployment scheduled for January 1986.

29 April–4 May: Aircraft carrier qualifications in the Southern California operating area abroad *Enterprise*. These operations were for the purpose of training in a feature found on the ICAP 2 machines, namely the Mode One Automatic Carrier Landing System (ALCS). A Mode One approach is a 'hands off' approach all the way to touchdown on deck, flown by a computer onboard the aircraft.

6 May–17 May: Air wing combat training at NAS Fallon, Nevada. One aircraft made the trip to Fallon, once one of the loneliest places in the world (in 1961, when the author spent several years there on an overnight stop, it was still called a Naval *Auxiliary* Air Station) and now, with its vast open spaces, an ideal spot to test for battle. The 12-day detachment to Fallon gave the Wizards their first chance to employ the ICAP 2 Prowler in coordinated air wing strikes (Alpha Strikes), using the ICAP 2 variant's new Inertial Navigation System, geographic displays, and improved jamming pods.

TOP LEFT
Two crewmembers inspect their EA-6B Prowler prior to a mission over the Atlantic from the USS John F Kennedy *(CV-67) in March 1978. The aircraft belongs to VAQ-133 'Wizards', assigned to Carrier Air Wing One (CVW-1)* (USN/Norman Polmar)

LEFT
A factory-fresh EA-6B, bureau number 161242, catches the soft light of a hazy day during a test flight from Calverton (Grumman)

5 May–7 July: at Whidbey. Training. It gets foggy and wet up there at Whidbey, home of all Prowlers and of west coast Intruders—a nice place to live, but for the lousy weather in which Prowlers and Intruders thrive.

8 July–22 July: refresher shipboard training on *The Big E*. One of the primary concerns was with the TEAMS (Tactical EA-6B Support Mission) computer which contains the information required to plan and execute an increasing variety of missions. VAQ-133 had the first TEAMS system designed for permanent installation on a carrier. While on *Enterprise*, the squadron received high marks for TEAMS' readiness and proper functioning.

23 July–28 July: back to Whidbey. Training.

29 July–8 August: further Southern California operations aboard *Enterprise*, this time for an exercise called COMPTUEX 1, the second workup period in preparation for the future Westpac deployment.

9 August–21 September: Whidbey. Training.

2 September–10 September: VAQ-133 sent two aircraft to MCAS Yuma, Arizona to participate in the Fleet Fighter Air Combat Readiness Program (FFARP). Flying with aggressor squadrons against aircraft from its own carrier air wing, VAQ-133 supplied ECM (electronic countermeasures) training to the fighter squadrons. This established a precedent whereby the EA-6B has become an integral part of FFARP exercises.

22 September–12 October: VAQ-133 returned to *The Big E* for COMPTUEX 2, another exercise readying for a Westpac cruise. This expedition gave squadron members a welcome four-day liberty call (holiday) in San Diego.

13 October–27 October. Whidbey. Training.

28 October–25 November: back to 'the boat' again for an extended Operational Readiness Exercise (ORE), further preparation for Westpac, with the Wizards attaining a score of 'excellent'.

26 November–31 December: in time to celebrate American Thanksgiving (the third Thursday in November), the Prowler squadron again returned to Whidbey. Though the history ends at this juncture, soon afterwards the Prowlers were aboard *Enterprise* again for the extended Westpac deployment which is the stuff of carrier aviation.

TOP RIGHT
*A pair of EA-6Bs from VAQ-137 'Rooks' use their wingtip airbrakes to hold station on the camera ship. This is February 1975, when high-visibility markings helped to make the Prowler an even more interesting sight on a carrier deck (*Enterprise *in this case)*
(USN)

RIGHT
An absolutely immaculate EA-6B parked at NAS Leemore, California, on 5 October 1973 shortly after being delivered to VAQ-131 'Lancers'. An F-8J Crusader, bureau number 149186, hides in the background
(Duane O Kasulka)

If this sampling of Prowler activity gives any appearance of being dull or mundane—and, again, other squadrons are reminded that the Wizards received so much space first because they are typical and, second, because they put it on paper for future readers—readers like the author who spend their lives commuting to the office. Consider the separations, the sacrifice, and the intense spirit of dedication and talent which must go into a Prowler squadron's work-aday activities.

Yes, of course, there are other Prowler squadrons. The following year, the 'Scorpions' of VAQ-132, then under Cdr Skip Fincher, logged 2,102 flight hours and earned the Atlantic Fleet Battle E Award for operations aboard USS *Forrestal* (CV-59), enti-tling crews to paint a distinctive 'E' on the squadron's four airplanes. At that point, the Scorpions had gone nearly 17 years with 26,700 flight hours without a single flying accident. The squadron had also scored an outstanding record avoiding accidents caused by FOD (foreign object damage)—those gremlin wren-ches, ballpoint pens and headgear items which can, if not guarded, get sucked into an air intake and ruin an engine. Other units have scored similar records of high achievement in the all-important realm of operational safety.

Naval and Marine aviation is, in fact, remarkably safe in spite of the inherent risks of operating from carrier decks. The overall safety record of the EA-6B Prowler and A-6 Intruder, at least after the initial 1965 deployments, has been excellent and has gener-ally been more favourable than the record of other aircraft types. It cannot be emphasized too much that Prowler and Intruder people think of safety first, even in the pitch of combat, and the Navy publishes a flight safety magazine, *Approach*, in 1986 edited by Cdr Dave Parsons with help from the able Peter Mersky, which is widely distributed. Accidents don't happen often—but once in awhile it *does* become necessary for a Navy or Marine flier to conclude a flight dangling from a parachute. Even then, as this narrative will soon relate, aircrews stand an excellent chance of surviving the worst.

Over $100 million in close formation. Electronic warfare does not come cheap, but today no credible military force can afford to be without it. The 'Vikings' of VAQ-129 formate with a single machine from VAQ-132 'Scorpions' to produce a memorable publicity shot in July 1971
(Grumman)

Chapter 4
'Eject, Eject!'

The Marines belonging to the 'Bengals' of squadron VMA(AW)-224 took their aircraft aboard for the Western Pacific cruise of USS *Coral Sea* (CVA-43) in 1972 just in time to participate in the buildup of fighting leading towards the Linebacker campaign against North Vietnam. Less publicized was the fact that US combat crews were also flying missions in Laos.

Major Clyde D Smith, pilot of A-6A Intruder bureau number 155652, side number 505, callsign BENGAL 505, went into Laos on 9 April 1972 and came out the hard way.

Smith was toting bombs on a mission against the Ho Chi Minh Trail, the enemy's notorious infiltration route, when his Intruder was hit by a SAM (surface-to-air missile). His bombardier-navigator was killed. Smith made a safe ejection but was still dangling in mid-air from his parachute when the North Vietnamese saw him coming down and began to cordon off the area with red tape, expecting to isolate and capture the Marine pilot.

'It was hairy', says another airman familiar with the situation. 'They didn't get to shoot down an Intruder very often, and when they did get one, they usually didn't have the leisure to sit back and watch the pilot floating down into their midst. This time, they had the region where he would hit the ground roped off, the way you might put a perimeter around a construction site. They *knew* they were going to roll him up. *They wanted to get the rescue forces, too . . .*'

Marine squadron VMA(AW)-224 embarked aboard USS Coral Sea (CVA-43) for the carrier's 1972 combat cruise in North Vietnamese waters. On 8 March 1972, A-6A Intruder 155646, side number NL-501, call-sign BENGAL 501, leads a two-ship over the carrier. A month later, Major Clyde D Smith's BENGAL 505 was downed by a SAM over Laos
(USN)

Though Major Smith retained the use of his beeper, a hand-held radio which also provided voice contact, it did not seem likely that a rescue force could be mustered to reach him. Says Lt James R Casey, copilot of the HH-53C helicopter seeking to locate him, 'We had a chopper in the area but, at first, no escort to cover the chopper. We couldn't get fast-mover support . . .'

'Eject, eject!' (I)

Although Major Smith's aircraft had now become a tangled heap of twisted wreckage in the Laotian jungle, the story of the attempt to rescue the Marine pilot properly belongs in Intruder lore. Soon, almost everybody who could get into the air around him was trying to find a way to snatch up the stranded Marine. Among would-be rescuers was Captain Arnie Franklin, pilot of a Pave Nail OV-10A Bronco of the

23rd Tactical Air Support Squadron, operating out of Nakhon Phanom, Thailand. Franklin lived, ate and worked with the HH-53C helicopter and A-1 Skyraider crews who made up the rescue force. Because his OV-10A had greater night- and poor-weather capability than they did, Franklin often found himself trying to help with a rescue.

It soon became apparent that despite their early warning and efforts to cordon off the area, the North Vietnamese had not captured the Intruder pilot. In fact, Major Smith later told the *7th Air Force News* that he spent the night of 9–10 April in the forest well-hidden and 'lying awake listening to strange noises. One of the weirdest was a bird that made a sound just like the clicking hammer of a gun.' Unknown to Smith, he was in a poor spot to be picked up. The driven men of the rescue force including Casey, Franklin and others—and including Major James C Harding, the rather colourful and totally fearless

skipper of the A-1 Skyraider squadron—were trying desperately to communicate with Smith to warn him that he was in the wrong spot and tell him that he needed to move due east. The problem was, *how to get this message to Smith without getting it to the North Vietnamese, who monitored beeper communications?*

Rescue Efforts

Each aircrew member had filled out a 'survivor card' containing personal details about himself—to prevent the North Vietnamese from decoying rescue forces. Smith's included his social security number. This enabled rescue people to remain in touch with the downed Intruder pilot, but they could think of no effective way to explain the situation to him: Smith believed he was safe because he was not being bothered. (Later, he would say that he'd seen the enemy walking by 'quite casually—they didn't seem too interested in finding me'. Major Smith was unaware that the North Vietnamese knew *exactly* where he was. The enemy was trying to lure in, and

kill or capture, the OV-10A, A-1, HH-53C and A-4 Skyhawk crews in the rescue force.

During the second day, Smith temporarily lost contact. He had no food and water, and gained some bodily replenishment only when a brief rain shower drenched him. He was still unaware that the North Vietnamese were trying to sucker rescue forces. It was unusual for anybody to be down this long without being captured, but Smith hunkered down for his second night in enemy terrain—wearied, hungry, and unaware of efforts to move him eastward.

On the third day, Pave Nail OV-10A pilot Arnie Franklin thought he had found a solution. Mindful that Hanoi's troops were listening in, Franklin penetrated the heavy cloud cover, spiralled down to the hill where Smith was situated, and sent a message for Smith to listen for the peculiar sound of OV-10A propellers overhead. His efforts were briefly interrupted when North Vietnamese gunfire struck his centreline fuel tank, forcing him to jettison it. Under the weather, *sans* centreline fuel, thinking only that 'we needed to *move* him and move him now!' Arnie

LEFT
Flying the Intruder is enough of a challenge, even when you don't have to eject from it. An oxygen-masked A-6A Intruder pilot maintains careful formation with a wingman, barely visible in background, during a carrier cruise
(via Dave Ostrowski)

BELOW
An A-6E dries out on the wash rack at the Naval Air Test Center at Pax River in Maryland on 5 October 1983
(Joseph G Handelman, DDS)

Franklin sent Intruder pilot Smith—and the eaves-dropping North Vietnamese—the following message:

'We think you need a cat before we can give you your trap. The wind is the first three of your SSN. Cyclic ops is now. Do you understand?' Back from Smith came a response saying yes, he *did* understand. But the OV-10A pilot was being too clever by half. In fact, Major Smith did not get the meaning of the coded message, which translates as follows:

A cat (catapult) is a launch and a trap is shipboard terminology for a safe landing. Franklin wanted Smith to 'launch' to a new location before he could give him a rescue helicopter. The first three digits of Smith's social security number happened to be 090—due east, the direction Franklin wanted him to move in, a significance the listening North Vietnamese would not be able to grasp. The final sentence was intended to tell Smith to move immediately, rather than waiting.

Major Smith, however, did not move east.

The third day faded into darkness without a rescue. No man had ever been down for this long in the vicinity of so many North Vietnamese troops.

On the fourth day, the North Vietnamese seemed ready to give up their attempt to snare the rescuers. HH-53C helicopters from the 40th Aerospace Rescue and Recovery Squadron (ARRS) orbited above the cloud-decked Laotian terrain, the lead ship piloted by Captain Bennie D Orrell with Jim Casey in the co-pilot's seat and Pararescuemen A1C William T Liles and Sgt William D Brinson manning the rescue hoist and miniguns. Major Harding's Skyraiders and Captain Franklin's OV-10As also pirouetted over-head, confounded by the weather but aware that the North Vietnamese were now within a few hundred yards of their man. Two US Navy A-4E Skyhawks from squadron VA-55 aboard USS *Hancock* (CVA-14) piloted by Lts Ken Bray and Tom Latendresse proceeded to use AGM-45A Shrike anti-radiation missiles to attack SAM sites nearby. Harding, Orrell and Casey decided they could wait no longer . . .

Harding, described by another pilot as 'having ice water in his veins', kept his Skyraiders at treetop level, working over the North Vietnamese with CBU-24 cluster bombs and 20-mm gunfire. Orrell gained eye contact with Major Smith on the ground and brought in the HH-53C. Casey, in the right seat, recalls that 'When we went into a hover to pick him up, it was like a World War 2 movie. We saw the bad guys coming at us. Their guns were firing. We had ordnance [from Harding's Skyraiders] so close that it damaged the sponson of the HH-53C.' A sidelight to the story is that a second HH-53C, callsign JOLLY 2, ran into problems when the pilot became hypoxic climbing to 16,000 feet without oxygen. The

VA-128 'Golden Intruders' A-6E soaks up the sun at NAF Washington, DC, on 9 November 1985 (Joseph G Handelman, DDS)

rescue was successful, however: when the helicopter lowered its penetrator some distance away because of the heavy foliage, the Marine major 'came busting through the woods like a fullback, grabbed the penetrator, and hooked it to his parachute hoist'. Major Clyde D Smith was literally hoisted to safety by his own D-ring and—for the pilot of one A-6A Intruder, anyway—near-capture by the North Vietnamese was transformed into a successful rescue.

It was not always so. As noted elsewhere, A-6 Intruder losses in Southeast Asia were higher than they should have been. The US Navy lost 51 Intruders in combat, 12 to operational causes. The Marine Corps sustained 16 Intruder losses (15 A-6A, one EA-6A), had two more Intruders destroyed on the ground during separate mortar attacks on Da Nang, and had four more claimed by operational causes. Particularly during the first combat deployments, some Intruders were lost to prematurely exploding bombs—a sign of the difficulties with bomb fuses which were never fully resolved.

'Eject, eject!' (II)

On 12 May 1974, bombardier-navigator Lt Robert

An A-6E Intruder from VMA(AW)-332 'Polkadots' shares the ramp at NAF Washington, DC, with a T-34C Turbo Mentor and a TA-4F Skyhawk on 5 February 1983 (Joseph G Handelman, DDS)

V McHale found himself flying through a warm night off the shores of Hawaii aboard a KA-6D, bureau number 152637, side number NE-523, callsign ELECTRON, belonging to the 'Main Battery' of VA-196—flying from USS *Ranger* (CVA-61). (McHale, unlike the airframe, belonged to the 'Swordsmen' of VA-145 skippered by Cdr Fred Metz.) The KA-6D, the Intruder's tanker version, was still equipped with the Martin-Baker GRU-5 seat, considered effective at zero altitude but requiring 100 knots of speed for a safe ejection. (All KA-6D airplanes in the Fleet have since been retrofitted with the zero-zero GRU-7.)

They were refuelling F-4 Phantoms. They completed the linkup and departed to the right. Tanker control vectored the KA-6D to the ship. It was 2130 hours. There was no moon, no horizon. McHale's pilot, Lt John Adams, apparently did not see what McHale saw: lights of the nearby F-4s, not where they were supposed to be, and dangerously close. (McHale's recollection is that at this time, F-4 Phantoms had wing lights but were not equipped with the formation strip lights retrofitted later; if he is correct, *Ranger*'s would have to be among the last Phantoms to receive formation strip lights, seen elsewhere at least five years earlier.) 'The pilot really didn't see the F-4s . . .'

McHale did. He threw his arm up. The pilot continued to the right. The lead F-4 saw the Intruder on a collision course and pulled up. The second F-4 apparently did not see the collision coming and did

LEFT
*This A-6E belongs to the 'Bengals'
of VMA(AW)-224, pictured on
11 February 1984*
(Joseph G Handelman, DDS)

BOTTOM LEFT
*Another Marine Corps' Intruder
from the 'Polkadots' of
VMA(AW)-332 parked in front
of a VX-4 F-14A Tomcat. Bureau
number 155684, an A-6E TRAM,
was photographed on 9 June 1984
at NAF Washington, DC*
(Joseph G Handelman, DDS)

TOP RIGHT
*Equipped with multiple-ejector
bomb racks and long-range fuel
tanks, an A-6E TRAM of
VMAT(AW)-202 takes a break
at NAF Washington, DC on 23
November 1985*
(Joseph G Handelman, DDS)

MIDDLE RIGHT
*An A-6E of the Pacific Missile
Test Center (PMTC) seen during
a visit to Hill AFB, California in
November 1983*
(H J van Broekhuizen)

BOTTOM RIGHT
*A-6E TRAM 161685 of VA-85
'Black Falcons' rests with wing
slats and flaps deployed at its home
port of NAS Oceana, Virginia on
13 September 1986*
(Joseph G Handelman, DDS)

not react in time. The Intruder hit the F-4 between its canopy and tail with the upper surface of the Intruder's left wing.

The offending Phantom was F-4J, bureau number 158357, one of the very last J models off the production line. The Phantom, using the callsign CITY DESK, had been one of several planes from *Ranger's* carrier air wing manoeuvring in the Hawaiian night. McHale recalled flying near the Phantom but, 'I thought we'd missed him'.

It was important to be careful. Vortices from the wingtips of a high-performance airplane can create a 'tornado', a force stronger than that of jet engine exhaust, capable of ripping a wing or control surface to shreds. Is this what happened? McHale doesn't know. He remembers only a bright flash.

'It was very bright inside the cockpit. There was a bright flash, then an explosion. [The blast was caused by his own pilot ejecting, but McHale didn't know it at the time.] Bright light. The airplane was gyrating, a piece of the wing gone. The ejection handle is down between your legs, under the crotch. I found the handle and went out. My ejection made a sound like . . . THUPPPP!!!'

Bob McHale didn't know it but he had a broken neck and two broken C-2 vertebrae. He recovered consciousness dangling beneath the canopy of an open parachute. 'I was floating like a dream. I was thinking, "What a nice Hawaiian night". My mask was hanging, not fastened. My gloves were in the G-suit. I was looking for my strobe light in my vest and . . . just then, I hit the water.'

McHale felt a sudden fear that his chute would drag him under the warm, churning sea. His right hand was burned. He reached for his Koch fasteners to disconnect the left. This seemed worse. When he reached up, he couldn't feel anything. He quickly thought of his gloves, worked them on, and made the disconnect. He paddled away. He had not had to use one of the many tools available to naval fliers, a shroud-cutter which is a U-shaped flat blade with a plastic handle, the inside of the U being the blade handle.

Now, thought McHale, as yet unaware of the seriousness of his injury, *what about getting rescued?*

He heard the helicopter searching. The helo was an SH-3 from the HC squadron aboard *Ranger*, its pilot Lt John Winters. 'The F-4 crew had made a textbook, no-problem ejection and were picked up by another helo.' Winters picked up McHale's pilot first, 'then he dragged his horse collar across the water and hoisted me up'. When first lifted from the water, McHale felt a tremendous weight pulling on him. He remembered that he'd been trained not to help himself into the helo, but to wait for the chopper's crew to assist. The weight confused him. 'I think I have my parachute attached—' The helo crewman looked at him and shook his head. 'No, sir. You're still attached to your seat pan.' McHale was heli-lifted to Tripler Army Hospital on Oahu.

Bob McHale's experience illustrates the fact that the Intruder is one of the few two-seat aircraft where the two ejection seats operate independently of each other. Thus, it's possible, as happened to McHale, for the BN to look to his left and see that the pilot has already left the aircraft. McHale says his pilot later insisted that he'd shouted, 'Eject, eject!' before leaving the Intruder—but McHale never heard him. Bob McHale recovered fully from his injuries, returned to flying, and more recently was assigned to the US Navy's European headquarters in England.

'Eject, Eject!' (III)

On 11 November 1985, the 'Green Knights' of Marine attack squadron VMA(AW)-121 'chopped' to operational control of the US Navy for a four-year period, becoming an integral part of carrier air wing two aboard USS *Ranger* (CV-61). The wing is an 'all-Intruder' air wing, lacking the A-7E Corsairs or F/A-18A Hornets found on most shipboard deployments.

The squadron has a proud history, too. As VMF-121, it produced many Marine aces during World War 2, including Major Joseph L Foss, who won the Medal of Honor. Designated VMA-121 during the 1950–53 Korean conflict, the 'Green Knights' flew AD Skyraiders from rough, forward airfields. In 1961, the squadron acquired A-4 Skyhawks, which it later operated from Chu Lai, South Vietnam. In 1969, the A-6A Intruder became the squadron's aircraft and the VMA(AW)-121 designation was created to signify the all-weather role. The A-6E model followed in the 1970s.

On 28 June 1986, the squadron was at sea aboard *Ranger*. Captain Vincent Paul (Brillo) Higgins woke up at 2:30 am, showered, and proceeded to the chow hall for two glasses of juice before being briefed for a pre-dawn launch. In the ready room, Higgins and his bombardier-navigator, Captain Richard Racine, studied the weather and their tactical assignment on closed-circuit TV. They were given a surface surveillance mission with a 5:00 am launch. Higgins and his B/N went through the usual drill for what was as close to a routine mission as any ever gets aboard a busy carrier: they conducted a thorough brief from the NATOPS briefing guide, covering thorough communications, navigation, rules of engagement (ROE) for surface contacts, fuel management, and emergency procedures.

Into the Mission

On completion of the brief, Higgins and his BN examined the maintenance records of their aircraft, A-6E Intruder, side number 400, modex code NE, bureau number 159902. On this mission, 159902 would carry two 300-gal drop tanks on ordnance stations one and five, a D704 'Buddy store' on station three, and multiple ejector bomb racks (MER) on stations two and four. The crew's primary mission

A-6A Intruders dropping bombs during a mission over Vietnam
(USN)

would be to observe and photograph surface vessels operating near *Ranger*. Satisfied from records that 159902 was safe for flight, the pair donned their flight gear and walked to the aircraft. The Intruder was parked on *Ranger*'s bow. It was approximately one hour until sunrise.

Captain Higgins conducted a walk-around preflight inspection of the A-6E and found no discrepancies. According to custom, he wielded his flashlights with the red lens off the light to make any leaking hydraulic fuel visible.

Higgins and Racine strapped in and were towed to the area aft of Catapult 1 for their start. Higgins noticed some minor difficulty with system alignment, but the aircraft was soon ready to go and he taxied to Cat 1 for launch.

As mentioned earlier in this narrative, the A-6 Intruder introduced the method of launching carrier aircraft using a catapult launch arm protruding from the nose wheel, a new concept when first employed but now routine. Higgins' pre-launch engine run-up and check of flight controls was normal. He turned off the aircraft exterior lights and gave the final signal

for launch. *Ranger*'s powerful steam catapult sent the Intruder hurtling into the dawn.

Higgins and Racine proceeded to their search sector and carried out a routine surface surveillance mission, investigating and photographing numerous small vessels. On completion of the mission, Higgins established a high power setting to return to *Ranger* (about 92 per cent rpm). Higgins recalls that the A-6E was flying well, with some minor discrepancies unrelated to the problem which followed.

'We were 2,000 lbs ahead on our fuel calculations and hoped to join on the overhead [KA-6D] tanker for dry plugs and to give him our extra fuel [using the buddy store]. We were level at 10,000 inbound on about 030 radial off [*Ranger*] when we lost the right flight hydraulic pump. The indicator was at zero. The time was approximately 6:50 am.' Higgins customarily scanned his instruments constantly, and felt he'd noticed the failure almost instantly. Things were no longer routine: with a limited hydraulic failure, an A-6E *might* be able to land with 'everything hanging down' in a 'dirty' configuration, but nothing was assured.

And if the other hydraulic pump should go . . .

Suddenly, the churning seas beneath the Intruder were very real indeed. Higgins got a squadron officer on the radio. The advice was to 'dirty up' the Intruder and attempt a landing. Higgins lowered the gear and got indications of three wheels down and locked.

89

Then he lowered slats and flaps. He was over the ship now, and had begun a descending 360-degree turn outside of ten miles to a holding altitude of 4,000 feet in anticipation that a recovery would be possible. Higgins established position and prepared to watch other *Ranger* aircraft land first, clearing the way for him.

At 7:02 am, two minutes past Higgins' originally scheduled recovery, things got considerably worse.

Trouble Aloft

'Five to six minutes after dirtying up, our remaining flight hydraulic pump, the left pump, went to zero. The backup hydraulic light came on along with the flashing master caution light. The failure, just like the initial failure of the right pump, was not accompanied by any noises such as banging or whining . . .' Higgins alerted the carrier. Over the airwaves, a VMA(AW)-121 squadron officer verified his understanding that the aircraft had experienced a complete flight hydraulic failure.

The A-6E was heaving along, flying admirably Higgins says. No sooner had he grasped the more serious failure than the left flight pump came back on the line to indicate 3,000 psi. The pump had been indicating zero for a total of three or four minutes. Higgins alerted the carrier to the fact that the left flight pump had come back up.

At this juncture, Higgins believes too many people were talking at once. 'By this time, the scheduled 0700 recovery was getting underway. There appeared to be some question of . . . recovery procedures. Certain aircraft received marshall instructions and certain aircraft were told merely to descend for the break, to the best of my recollection. The delay was compounded . . . by discussion of assignments and instructions between various aircrew and the duty controllers aboard. After a period of about five minutes of holding [during which time Higgins seems to believe it would have been possible for him to attempt a landing], our remaining flight hydraulic pump, the left pump, once again failed as our indicator dropped from 3,000 psi to zero. Once again, we radioed an update of our hydraulic situation to our squadron representative aboard *Ranger*. The time was now about 0715. We would remain without hydraulics . . .' The Intruder was now slewing through the wet morning flying only on backup hydraulics. The backup hydraulics had been designed solely for the purpose of flying a battle-damaged aircraft away from a hostile environment prior to ejection, and Higgins quickly learned that the

The 'Thunderbolts' of VA-176 are based at NAS Oceana, Virginia where this study of an A-6E on the line was taken on 11 October 1981
(Don Linn)

backup was not good for much else. 'Control of the aircraft degraded rapidly . . .'

After some manoeuvring and watching *Ranger* emerge into better weather, Higgins completed a turn and received instructions to proceed outboard to get into position for a straight-in approach following a Lockheed S-3A Viking, side number NE-702, which was experiencing a vibration problem. Higgins initiated a descent at 150 knots. He began a gradual left turn downwind while keeping throttle movements to a minimum. But it was becoming almost impossible to handle the Intruder.

Lateral controllability was only available through rudders, as the flaperons are inoperable in the backup system. Through smooth operation of rudders, I was able to maintain an approximate wings-level attitude (plus 10 degrees angle of bank on either side). I was able to execute a 90-degree left turn and roll out on a base leg to hopefully intercept the final bearing. I initiated this turn at six miles when we were cleared to land ahead of the other aircraft with problems.

A Marine Corp's A-6E Intruder from VMA(AW)-533 'Hawks' makes a copybook trap aboard a carrier in the Sea of Japan
(USMC)

'A more serious control problem than the lateral axis was the difficulty in setting and maintaining nose attitude on the longitudinal axis. The control stick had a different "feel" throughout its fore and aft range. There was a dead spot in the centre of the stick. Slight backstick required normal pressure. However, the nose attitude response was delayed and erratic. The full back stick area witnessed extremely difficult stick movement. Both hands were required to ease the stick full aft and forward once again from the full aft position. The forward stick area had equally inconsistent feel, with a small quadrant of "easy" motion and a large range further forward with difficult stick movement.'

An EA-6B Prowler catches the second arrestor gear cable aboard Nimitz *at the end of mission during Exercise Display Determination on 14 October 1981. The motif on the rudder makes it clear that this is a 'Playboy' from 'Q-2'* (USN/PH1 Douglas Tesner)

Out of Control

Higgins was having enormous difficulty exercising any control over the Intruder. He was getting a delay of 5 seconds between stick input and a corresponding change in nose attitude. With a struggle, he began to get the hang of the required corrections. For an increase in nose attitude, the stick would have to be smoothly eased aft through the difficult area of movement. He learned to hold the stick there for a moment and began moving it back through the stiff area towards the null position. As he began to move the stick back the lag would catch and the nose would come up. The slow nose movement was never more extreme than about 5 degrees nose up or 5 degrees

down throughout Higgins' battle with the crippled aircraft.

'The path of the aircraft on downwind and base was generally a shallow descent. There were two occasions that I remember levelling off and even slightly climbing. I fully realized the aircraft was in a shallow descent. The stiff region of backstick authority made it very uncomfortable to climb. Occasionally the delay and subsequent rising nose would put the aircraft a couple of units slow. Full power on the engines (which functioned perfectly) and an easing forward of the stick would arrest the "slow".'

Higgins was trying to buy time. 'Flying [a shallow descent] allowed us time to prepare for ejection and kept us out of the stall regime. I did not intend to stall the aircraft and jeopardize our ejection envelopes.' In the adjacent seat, Racine asked Higgins several times if he had control of the aircraft. Commander Roach, the landing signal officer (LSO) on *Ranger*, asked the same thing. Higgins responded

TOP LEFT
*An EA-6B Prowler from VAQ-131 'Lancers' launches from
the deck of* Independence *(CV-62) during a Mediterranean
cruise in which the carrier operated in support of the multi-
national peacekeeping force in Beirut, Lebanon, in 1983*
(USN)

LEFT
*The 'Black Panthers' of VA-35 form an impressive line up
at NAS Oceana, Virginia in August 1979. A KA-6D tanker
is parked in the middle*
(USN)

ABOVE
*Safely tied down, an A-6E TRAM from VA-34 'Blue
Blasters' hugs the edge of the deck to maximize parking space*
(Dave Parsons)

LEFT
The 'double nuts' side number identifies this A-6A Intruder as the CAG's personal aircraft aboard the Air Wing on Saratoga (CV-60). This portrait of an Intruder belonging to the 'Sunday Punchers' of VA-75 was probably taken in the early seventies
(USN)

BOTTOM LEFT
Shortly after the Western Pacific cruise in which he had to eject from an Intruder, Captain Paul Higgins and his squadron mates of VMA(AW)-121 'Moonlighters' were working out in the desert. A-6E TRAM Intruder side number 415, its bureau number not visible, is dropping a full load of Mark 82 Snakeye retarded bombs at the B-17 East Target Complex near East Fallon, Nevada. Crew is 1st Lieutenant Tom (Mud) Bresnahan and Captain Jim (Man of Steel) Vile
(Paul Higgins)

Captain Paul Higgins, USMC, seen just before the Western Pacific cruise by VMA(AW)-121 during which he had to eject from an Intruder
(courtesy of Capt Higgins)

Despite the fact that three crewmembers seem intent on strapping themselves into this A-6E, there's only room for two in the Intruder's cockpit. The bombardier-navigator (BN) sits on the right, his console dominated by the radar scope directly in front. This Intruder formed part of CVW-9 onboard the USS Constellation *(CV-64) when this picture was taken on 19 April 1980* (USN)

that, yes, he barely had control—but he and Racine exchanged glances in the full knowledge that they were going to have to punch out, leaving the Intruder in the early dawn daylight with an unfriendly ocean below.

Racine told Higgins of his intention to eject approaching 800 feet. Higgins told the bombardier-navigator to take off his knee pad and get ready. Racine lowered both of the Martin Baker GRU-7 ejection seats in unison to lessen the danger of neck injuries. In his earphones, Higgins heard the LSO directing *Ranger*'s search and rescue chopper 'down the wake' and advising Higgins not to stay with the Intruder too far into the continuing descent.

Higgins pushed both handles to military power and told Racine to get out. He grabbed the stick with both hands and eased forward as the nose was coming up and he was watching angle of attack very carefully. At the time of ejection, he was holding the wings level with right rudder. 'The instant I saw Racine's seat flash and heard the wind noise, I held the stick forward with my left hand and pulled the lower ejection handle with my right hand.' As far as can be determined, the ejection took place at 400 feet altitude and 130 knots at 7:25 am.

'I saw and felt myself tumbling end over end in the seat. I vaguely remember seat/man separation and stabilizing with the drogue chute. My main canopy opened with a shock, jerking me slightly backwards. I inspected my canopy, raised my visor, unclipped my mask (which was still supplying oxygen) and relaxed. I made a quick decision not to employ the four-line release because I didn't think it necessary.' Higgins may have been in the first flush of relief at being alive, having not yet learned that his own A-6E Intruder intended to make one more attempt, still, to kill him.

'Suddenly, to my surprise, I saw the airplane. NE-400 was climbing straight up 250 feet away presenting me with a platform view of the top of the aircraft, I could see the numbers on the right inboard flap, dirt and bootmarks on the paint. I saw no sign of pink hydraulic fluid. The engine noise made it clear that both throttles were at military. Not knowing if the airplane had the energy to continue into a loop and kill me, I quickly changed my plan with the four-line release decision. I glanced up and did not immediately find [the parachute release lines]. I looked back at the airplane and quickly realized from its path that I was out of danger . . .'

During his fifteen-second parachute ride, Captain Higgins attempted to locate his bombardier-navigator. 'I spent a few moments trying to deploy my seat pan with no success. I could not locate the handle. I grabbed my Koch fittings to prepare to hit the water. As my feet hit, I undid both fittings and got out of the parachute.'

A couple of shroud lines were wrapped around Higgins' leg restraint. He quickly cast them off. Higgins looked around for the BN and yelled out. He

A drab EA-6B Prowler flanked by a rather more upbeat A-7E Corsair of VA-22 'Sidewinders' (left) and a S-3A Viking anti-submarine aircraft aboard Nimitz *during preparations for a mission in support of Exercise Display Determination with NATO allies on 19 September 1981* (USN/PH1 Douglas Tesner)

heard Racine yell back and saw him a couple of times on the crest of a wave. The BN was all right. 'I attempted for the next five minutes to deploy my seat pan. Although it sounds ridiculous, I could not find the handle. I wonder if the handle did not separate from the seat pan during the ejection. I felt something like a stub where I thought the handle should be but I'm not sure. Hearing the SAR helo approach, I undid the fittings on my torso harness and released the seat pan.'

As the SH-3G helicopter approached, Captain

Higgins waved and gave a thumbs-up. He gestured to the helicopter pilot to pick up Captain Racine first. 'I felt in perfect health and was comfortable in the 72-degree water.' Higgins watched Racine's uneventful rescue. The SH-3G then hovered above him and a rescue swimmer dropped into the sea at his side. 'I told the swimmer I was OK. He hooked us up to the hoist, which lifted us into the H-3. The return flight to the ship was uneventful. My only "injuries" were my sore right calf and a very minor scratch on the left index finger.'

Though the safety record of the A-6 Intruder is remarkably good, aerial warfare and aircraft carrier operations pose risk which is a part of everyday life. Smith, McHale and Higgins learned that every takeoff isn't necessarily followed by a landing. The unexpected in aviation is, of course, precisely what makes it so interesting. These incidents do not, however, detract from the fact that the A-6 is a thoroughly reliable machine in which air crews place enormous trust.

Walk-Around Check

Viewed from almost any angle, the operational A-6E TRAM Intruder attack aircraft seems almost to invite derision from anyone who thinks airplanes should be beautiful. The low-visibility paint scheme adopted by the US Navy in the early 1980s—a defence against not merely the enemy's Mark One Eyeball, but also his infrared heat-seeking weaponry—makes the Intruder appear drab and colourless. Indeed, the paint job is so dull that side numbers have had to be painted in black, so that they would stand out enough to allow carrier deck hands to identify individual aeroplanes.

If we begin our walk-around check the way many pilots do, by walking out to the wingtip and taking a hard squint at our flying machine, we will see a wing which is swept back almost 45 degrees at the root for only about three feet (one metre) to quarter-chord where the wing assumes its lesser sweep angle of about 25 degrees. The wing root has an upper non-slip walkway enabling air and ground crews to gain access to the rear cockpit area and upper fuselage. If we could look at the wing itself from directly above we might, at first, wonder if this airplane were a Soviet product from an OKB (design bureau). Far from eluding the essence of sleekness and streamlining, the wing is a mottle of control surfaces, tip brakes, fences, fold hinges lines and walkways. The air brakes which can slow down the Intruder very dramatically were, of course, re-positioned from the fuselage to the wingtip (except on the EA-6A model), possibly because of a fatal accident at NATC Patuxent River, Maryland, in the early 1960s resulting from the pilot *not knowing* his brakes were extended!

An intriguing sidelight here is that one of the other designs submitted for the competition which produced the original A2F-1 Intruder was rejected by the US Navy precisely because it had wingtip brakes.

The wingtip air brakes require massive hinge fairings. Radar warning receiver (RWR) antennas are located in fairings beneath the wingtip and along the leading edge of the wing glove. A ram-air turbine (RAT) capable of providing emergency electrical power in the event of a main systems failure is located at the wing root and is lifted outward by an access panel door. To the naked eye, the wing of the Intruder appears to be suffering from a benign skin cancer, its smoothness interrupted by ruffles and rumples. Wings of the A-6E TRAM Intruder span 53 ft 0 in (16.15 m), which is reduced to 25 ft 4 in (7.72 m) when folded and have an area of 528.9 square feet (49.1 m²).

Located at the wing roots are the powerplants for the Intruder, two non-augmented 9,300-lb (4128-kg) static thrust Pratt & Whitney J52-P-8A turbojet engines. This medium-sized turbojet engine designed and developed under the auspices of the US Navy's Bureau of Weapons has never really received the accolades it deserves, even though the engine has had a long and trouble-free career powering the A-4E

and TA-4J Skyhawk, the AGM-28 Hound Dog missile, and all Intruder variants with the exception of the forthcoming A-6F (of which, more soon). A two-spool turbojet with twelve compressor stages, the J52 employs cannular combustion, fed by 35 dual-orifice injectors and mechanically independent high and low-pressure turbines. In plain English, it is a practical and workmanlike powerplant for the Intruder.

As we continue walking around our A-6E TRAM Intruder we observe a number of obvious features which have already been commented upon in this text, ranging from the bulbous nose to the all-moving tailplane and the blister which houses an electronic antenna in the rear part of the fin. The Intruder looks a little as if it is planning to impale some hapless victim on the tip of its bent in-flight refuelling (IFR) probe which is not retractable (it can be removed, but rarely is) and appears unduly large. In fact the IFR probe is a handy way of 'aiming' the airplane, as well as refuelling, and pilots insist that it does not hinder their vision.

The fuselage of the A-6E TRAM Intruder, as indicated in our earlier reference to the basic design, is of conventional all-metal semi-monocoque structure, the bottom being recessed between the engines to provide space for a centreline ordnance or fuel store. The hydraulically retractable, tricycle landing gear includes the twin-wheel nose unit which retracts backwards and single main wheel units which retract forward and inward into air intake fairings. An A-frame tailhook for carrier landings is located under the rear fuselage. A considerable amount of the 'working' maintenance area of the aircraft, including the all-important engine bays, is relatively low to the ground (or carrier deck) and therefore accessible without elaborate ladders or dollies. The 'birdcage' in the lower rear area of the fuselage which holds a variety of equipment, is actually large enough for a maintenance man to fit inside, and can be lowered while the aircraft is on the ground. The fuselage is 54 ft 7 in (16.64 m) in length, a figure which is increased to 55 ft 6 in with the EA-6A variant.

The A-6E TRAM Intruder crew's two Martin-Baker GRU-7 ejection seats can be reclined to reduce fatigue during labour-intensive terrain-following missions. The bombardier-navigator's seat is slightly behind and below the pilot to the right. In an emergency, it is possible to eject through the hydraulically-operated, rearward-sliding canopy.

On balance then, our Intruder may not be especially glamorous but it is a sensible and practical machine. If it is not pretty, it reminds us how well men build machines.

Senior chief aviation machinist's mate Joseph A Lang, back to camera, the maintenance control chief of VA-65 'Tigers', confers with the crew of an A-6E as final adjustments are made to the aircraft prior to launch from the USS Independence *in October 1974* (USN/JOCS Dick Benjamin)

Chapter 5
Combat: Lebanon

For a few tense weeks in November 1983, American naval aviators believed they were about to inflict a punishing air strike against Syrian troops and terrorist bastions in Lebanon. They were ready. They had a plan. They had the means.

Above *all*, they had the means to achieve what seemed easily achievable while they were preparing and fine-tuning the plan: they could, and would, inflict a blow so devasting that neither Islamic Jihad nor anyone else would mess with the United States for a long time to come. When the plan unravelled—when fanatics forced the world's most powerful nation to retreat from Lebanon—the beliefs of those Navy fliers may have been altered. But if there was fault to be attributed, none of it belonged to the Grumman A-6E TRAM Intruder and there never existed a moment's doubt that, used properly, the men and machines could have done the job. Once again, old men who sit at desks must be blamed for what went wrong while young men in Intruder cockpits should be credited for anything that went right. The A-6E TRAM Intruder, almost ludicrous in appearance with its tadpole shape and yet as deadly as a pit viper (to mix metaphors) lay at the centre of the top-secret American plan to reinforce a presence in Lebanon at the very time when critics were saying that no way existed to maintain that presence, no matter what.

US naval aviators, including A-6E TRAM Intruder crews, were inexperienced fighting against state-of-the-art weapons but the Middle East had long been a testing ground for exactly such hardware, among the world's most advanced. In its continuing difference of opinion with Syria, the state of Israel kept confronting modern field-sweep radars, mobile SAM missiles, and late-model MiGs. In fact, Israel's F-15A Eagles shot down their first MiG-25 *Foxbat* on 13 February 1981 and downed a pair of MiG-23 *Floggers* on 2 May 1982. As early as September 1983, with the nuclear-powered USS *Dwight D Eisenhower*

(CVN-69) on alert at Bagel Station off the coast of Lebanon, naval aviators were evading gunfire while carrying out reconnaissance missions. So far, none of the naval aviators had gotten hit, although some Syrian 37-mm cannon fire did come perilously close to a TARPS-equipped (reconnaissance) F-14A Tomcat of 'The World Famous Pukin Dogs' of VF-143 in the Chouf Mountains. *Eisenhower*'s Intruder and Corsair pilots wanted to retaliate, but although some form of retaliation was very much on American minds, it would not happen until *Ike*'s period on the line came to an end.

In 1982–83, the US sought to shore up the fractured government of President Amin Gemayel by putting Marines on the ground in Beirut and assembling an armada off the Lebanese coast. The American presence, it was hoped, would bring stability to Lebanon following the disruptions which accompanied the Israeli invasion of 1982, the increased intervention of Syria, and the ensuing violence between Lebanese factions. Too, the American presence would deter the extreme fanatics—Islamic Jihad, one group called itself, declaring a Holy War—allowing moderates to dominate the political scene in Lebanon.

In 1983, a suicide-bent terrorist undercut this show of strength by driving a truck filled with explosives into the US embassy in Beirut, collapsing the chancery building and killing 29. A second bombing followed. It was a devasting setback, but the real blow came on 23 October 1983 when a similar truck-bomb attack blew up the Marine billet at Beirut airport and, this time, no fewer than 241 men died. the repercussions affected almost everyone in the Marine corps, and real doubt arose as to whether the US presence could be maintained, let alone strengthened.

Adding to tensions in the region, on 20 November 1983 an Israeli IAI Kfir C.2 fighter was hit by Syrian gunfire. Its pilot ejected near the US Marine Corps billet and narrowly avoided becoming a prisoner.

At the decision-making level in Washington—meaning, almost certainly, inside Ronald Reagan's White House—a decision was made.

The US would regain its dominance in the region through a single, surgical air strike against terrorist installations and against the Syrian facilities which supported them, using the only carrier-based aircraft capable of such precision, namely the Grumman A-6E TRAM Intruder.

Washington Post staff writer George C Wilson was aboard USS *John F Kennedy* (CV-67) on Bagel Station at the time and describes in his book *Super Carrier* (New York: MacMillan, 1986) how the burden of preparation fell upon Rear Admiral Jerry O Tuttle. 'Sluf' Tuttle is an action-oriented leader in a Navy which has too many managers, too few leaders. Historian Hal Andrews calls him 'a really keyed-in decision maker who has a widespread reputation for getting the job done right'. To Wilson, Tuttle was 'brash, demanding, impossible, dangerous ... brilliant, warm, inspiring, futuristic'. Jerry Owen Tuttle's official biography notes his birth on 18 December 1934 in Hatfield, Indiana, and lists a dozen assignments, many in the attack community, including a stint as *Kennedy*'s captain (August 1976–November 1978). In more recent times, this holder of the Distinguished Flying Cross with two gold stars has become commander-in-chief of the US Atlantic fleet (since October 1985). If anybody could get an Intruder force into Lebanon, defy every defensive weapon from the AK-47 rifle to the SAM missile, and inflict precise, painful damage to the bad guys, Jerry Tuttle was the man you wanted at the helm.

Anti-Terrorist Mission

Although it turned out differently in the end, there was never any thought of employing any other aircraft type in the one-of-a-kind Lebanon air strike. It was to be an all-Intruder mission.

The 'Black Falcons' of VA-85 under Commander Kirby (Skip) Hughes and aboard *Kennedy* were to fly the strike. A mere eight Intruders, with precision-guided 1,000-lb (454-kg) Mk 84 bombs were to pulverize a terrorist redoubt in the Bekaa Valley. The 'Rooks' of VAQ-137 under Commander Bud Holl away would use their EA-6B Prowlers to jam and deceive the Syrian-manned radars in the area. Commander of the strike would be the CAG, or carrier air wing commander aboard *Big John*, Commander John J Mazach, who of course ranked the men in the Intruder and Prowler squadrons but was a fully-qualified pilot of both the F-14A Tomcat and the A-6E Intruder. The air strike would be short, simple and direct. Tuttle and Mazach were certain it could succeed without casualties. Commander Joe Prueher, *Eisenhower*'s CAG, was another of the key officers clued in on the secret plan.

Demanding professionalism, Tuttle told key officers that the first attack ever launched by the United States against an Arab country would be carried out at night, taking advantage of the Intruder's precision bombing potential. F-14 Tomcats from *Kennedy* would be aloft in case Syrian fighters ventured to intercept the strike force—not considered likely—and rescue helicopters would be alert not only aboard *Big John* and *Ike* but also from the HH-46D Sea Knight force aboard the helicopter assault ship USS *Guam* (LPH-9).

The *timing* of the mission was not known by anyone in the naval task force off Lebanon, not even Tuttle. Orders were issued to keep a number of A-6E TRAM Intruders 'bombed up' and on alert at all times. Meanwhile, although they did not engage forces on the ground in Lebanon, the task force's aircraft began reconnaissance missions and 'flag showing' flights over that troubled country, at times attracting fire. F-14, A-6 and A-7 pilots involved in this thrust-and-parry over a hostile land began to make their logbook entries in green ink, signifying combat.

While naval aviators were waiting for the decisive blow in Lebanon, the United States surprised everybody by invading Grenada. Operation Urgent Fury, the October 1983 seizure of that Carribean island nation in a US combined arms effort temporarily drew attention away from the Middle East. Other aircraft types aboard the carrier *Independence* (CV-62) received more attention, but the battle might not have been won but for the air strikes flown by the ten A-6E TRAM Intruders (and four KA-6D tankers) of the 'Thunderbolts' of VA-176 aboard *Indy*. Shortly after US Rangers seized Point Salines airfield, they encountered strong resistance from Cuban troops in tactically advantageous positions and had to call in air strikes from VA-176, as well as one of the A-7E Corsair squadrons aboard *Indy*. The Intruders helped again when the main assault on the town of St Georges was accomplished by helicopter-born Marines, supported by the A-6Es as well as A-7Es and AC-130H gunships. It was an air war in paradise as fliers from all the US services routed out resistance in a relatively tame air defence environment. No Intruders were lost. When compiling information for this volume, the author encountered some bitterness among Intruder crews that their contribution in Grenada had been overlooked: none was willing to be interviewed, however. The war in Grenada was fought without reporters present, relatively little material on the role of carrier aviation has been made available, and the Intruder crews themselves are mum. Under *those* circumstances, their contribution may have to be overlooked a little longer.

While attack pilots on *Independence* were about to face the unlikely eventuality of fighting in two wars on the same cruise, the secret plan for an air strike in the Middle East became rapidly unglued. French intervention, in the form of an ineffective 17 November 1983 air strike against Baalbek in the Bekaa Valley, apparently undertaken without consultations between Paris and Washington, gave unintended warning to the terrorist bases Tuttle was planning to attack. The carefully-nurtured plan for a small, effective, nocturnal Intruder strike began to unravel. Instead, as December arrived, the naval aviators found themselves continuing the thrust-and-parry while unable to take decisive action. The waiting was beginning to create tension, even among the highly professional Intruder crews who sensed that something was going to happen but didn't know what . . .

Intruder Man

The waiting must have seemed a challenge to Captain Gary F Wheatley, *Kennedy*'s captain, who was thoroughly briefed on Tuttle's plan and who knew it could succeed because he knew the Intruder. Annapolis graduate Wheatley had begun his naval aviator's career in attack squadron VA-44 in April 1961, and had met up with the Intruder at the west coast RAG, the 'Golden Intruders' of VA-128, in March 1968. Subsequent assignments with VA-145, VA-128 (again), VA-115 and VA-34 had made Wheatley one of the most qualified Intruder instructor pilots in the Fleet and he must have regretted that his seniority and responsibilities as *Big John*'s skipper made it impossible to actually fly on the big mission coming up—if, indeed, as everyone wondered, a big mission still *was* coming up. Born 11 October 1937 in Cleveland, Ohio, a soon-to-be rear admiral, Captain Gary Francis Wheatley now did the same thing as the other 5,000 men aboard his ship. He watched and waited . . .

On at least a couple of instances, A-7 Corsairs from *Eisenhower* took hostile fire over Lebanon and nobody did anything about it. (There were no Corsairs aboard *Kennedy*, which had an all-Intruder force.) Tuttle was alerted that the strike against terrorist-related targets, which had been 'off', was now 'on' again. According to some reports, a conflicting message informed the task group commander that he could stand down, as the strike would not be necessary after all. To strike or not to strike? Critics accused Washington of indecision. Tuttle, like his men, watched and waited . . .

Provocation

On 3 December 1983, an F-14A Tomcat piloted by Commander John (Market) Burch with Lieutenant John (Fozzie) Miller as back-seater went into Beirut with a TARPS reconnaissance pod slung under its belly, just as Tomcats from *Ike* and *Big John* had been doing for weeks. Burch, nicknamed for alleged skills with Wall Street securities, suddenly found themselves a bit insecure. Anti-aircraft guns began to pop off at them.

Carrier aircraft had drawn fire before. But now Commander Burch saw smoke trails creating quick, ugly little spirals against the ground below. These trails came from Soviet-built SA-7 man-portable missiles, IR (infrared) heat-seekers which could easily come rushing up the Tomcat's twin exhausts and blow the jet fighter to pieces. Burch completed his mission and returned to *Kennedy*. He knew he'd been shot at.

What he didn't know was, those SA-7s were going to draw retaliation. The air strike which had been so scrupulously planned in November was now resurrected, but now with a wholly different set of suppositions about when, where and how the strike should be carried out, with none of the meticulous planning that had come before, and without the same kind of preparation. To this day, no one knows why the decision was reached to fly one air strike only. To this day, no one knows why the strike was planned and executed as it was.

White House Orders

This time, the decision to attack was made by President Reagan on the evening of Saturday, 3 December, although the final go-ahead was given by Defense Secretary Caspar Weinberger who was visiting Paris at the time. All of a sudden, there was urgency— perhaps haste, as things turned out—in the decision to bomb a Syrian ammunition dump, a radar antenna dish, and surface-to-air missile (SAM) sites. The men who'd watched and waited were now to fly and fight . . .

Bobby Goodman

Annapolis graduate Lieutenant Mark (Doppler) Lange, 26, from Detroit, was typical of the bright, able, ambitious men in naval aviation and in Intruder cockpits. Not included among pilots selected for the original air strike in Lebanon, the secretly-planned strike which never took place, Lange talked his way into being one of the selectees when the 4 December 1983 attack was decided upon.

Ever after, it would be asked, again and again, why an aircraft with superb night- and bad-weather capability was sent into Lebanon in broad daylight, with the sun confronting the Navy aircrews and helping Syrian anti-aircraft gunners. It would be asked why the relatively slow Intruder was used under circumstances where the airplane's strong features had no opportunity to excel. The A-6E TRAM Intruder is something of an American national asset, being the *only* night- and bad-weather strike craft which (unlike the Air Force's F-111) does not raise sovereignty issues by requiring the use of some other nation's airfields. To be sure, the Intruder had been used repeatedly for daylight bombing in Vietnam, but was Vietnam the lesson planners should be using? The problem was not the decision to use the Intruder. The problem was the decision not to do it at night.

All of this may have been readily apparent later. On the morning of the mission, it was apparent only that the men who maintain, arm, and fly Intruders were extremely busy.

Like his squadron mates in VA-85, pilot Lange did not have a specific airframe assigned when preparations for the mission began in the Mediterranean darkness. With the approach of dawn, and launch scheduled for late morning, he still did not. Just as the strike was about to be mounted, at the last minute, Lange ended up with A-6E TRAM Intruder 152915, modex code AC, side number 556, one of several airframes aboard *Kennedy* which had begun life as an A-6A before being updated. For the mission, 152915 was 'bombed up' with a full load of six Mk 83 1,000-lb (454-kg) bombs. Lange's bombardier-navigator was to be an officer with whom he had not been paired recently, despite the Navy's strong emphasis on 'hard' crews.

Lieutenant Robert O (Bobby) Goodman Jr, 27, sometimes called Benny after the clarinet player, was from San Juan, Puerto Rico, the son of an Air Force KC-135 tanker pilot. Goodman was another very able graduate of the Naval Academy, his Annapolis ring a source of real pride. At first not fully aware that *Kennedy* was preparing for a short-notice attack, Goodman was awakened in his billet aboard the carrier at 4:30 am. Goodman was told—absolutely wrongly, it turned out—that launch was scheduled for 11:00 am. Like everyone, he had the 11:00 am deadline fixed in his mind. He was matched up with Lange, and told to be ready for 'the real thing'. At the White House, the decision had been made to retaliate against the Syrian gun positions which had fired on the Tomcat the day before. This was very decidedly *not* the nighttime anti-terrorist air strike into the Bekaa Valley which had been planned the month before and the change of plan posed a grave threat to Tuttle's prospect of commiting his forces without casualty. Tuttle's planning, Wheatley's professionalism, Mazach's leadership, and above all the mettle of Doppler Lange and Bobby Goodman, were to be challenged to an extent that no one had anticipated.

For reasons never explained, and beyond the control of Sluf Tuttle or his naval aviators, the White House—or possibly, Weinberger in Paris—made abrupt, last-minute changes to the plan for the strike. What had once been envisaged as a night attack by Intruders only was now to be a daylight assault by Intruders and A-7 Corsairs from USS *Independence* (CV-62), which had just left behind the brief war in Grenada, steamed into the Mediterranean, and relieved *Eisenhower*. At the *very* last minute what had been planned as a late-morning air strike was rescheduled for 7:00 am. Everybody from Tuttle down to Goodman was working frantically against the 11:00 am deadline when four hours' worth of planning and preparation were abruptly snatched away from them. The decision to launch earlier was reached with such stunning abruptness that many of the Intruders were only partly 'bombed up' and actu-

ally departed on the mission without a full load of bombs. At least one and possibly several of the Intruders went forth on the actual combat mission carrying only two small training bombs—there being no time to load proper ordnance. (The Corsairs and Intruders from *Independence*, comprising 18 aircraft of the 28-plane force and about which less information is available, may have been fully armed: one version of events has it that *Indy*'s Intruders were laden with twenty-eight 500-lb (227-kg) Mk 82 Snakeye fin-retarded bombs.)

Why 7:00 am, with the sun in the pilots' eyes? Why daylight? Why without enough bombs? Were the men aboard *Independence*, in fact, better prepared than those on *Kennedy*? Why did the United States pass up the opportunity to use the A-6E TRAM Intruder in its natural, nocturnal setting?

The changes in planning, particularly the decision to speed-up the morning launch, posed major difficulty for the two carrier air wings which were now to commit 28 Intruders and Corsairs to the strike. VA-85 skipper Hughes, a lean and lanky figure at six-feet-four (and a bombardier-navigator in a Navy which does not entrust command solely to pilots), did a superb job of preparing men and machines before climbing into an Intruder with Lieutenant Commander Mark (Mork) McNally. Bobby Goodman scrambled towards 152915 with both his flight gear and Doppler Lange's. To the uninitiated, it might have appeared that there was confusion on *Big John*'s deck as the sun came up and the men struggled to meet the abbreviated schedule. In fact, the only confusion existed on the part of policymakers who had, as they do so often, ordered men into battle while withholding the tools to do the job.

The attackers were striking out at a range of about 100 miles (161 km) carrying their emasculated ordnance toward their targets at a speed of about 400 knots (581 km/h). The distance was moderate enough. There existed real concern, however, that the carriers themselves, as well as the strike aircraft, might be threatened by the array of modern Syrian weapons on the other side.

CAG Mazach's force launched from *Kennedy*—some minutes behind the accompanying force from *Independence*—with incomplete bombloads and hasty instructions. An armada of supporting aircraft—E-2C Hawkeyes, tankers, rescue helicopters—also went aloft. No doubt the two carriers' F-14A Tomcat crews would have welcomed the chance to test themselves against fighter opposition. It was not to be, but a formidable array of other kinds of defences awaited the attack crews. Highly mobile SA-6 *Gainful* missiles, shoulder-mounted SA-7 *Strella* and SA-9 *Gaskin* missiles surrounded the targets. The attack aircraft almost certainly employed aids such as the Goodyear AN/ALE-39 chaff/flare dispenser to foil the accuracy of Syrian missiles.

The attack force itself bored toward the Lebanese coast. It had *actually happened*, though some of the men themselves could not understand why: the Lebanon air strike had been set under way with the men flying into the blinding red glare of the rising sun.

Into the Assault

Some reports of the raid have indicated that Intruders aborted with mechanical problems. One version of events holds that one A-6E TRAM Intruder was hit by gunfire the moment the strike force crossed the beach, peeled away trailing smoke, and successfully recovered on its carrier—not to be counted among the casualties for the day.

Mazach's Intruders employed their TRAM to acquire gun positions and munitions sites manned by the Syrians at Falouga and Hammana, ten miles (1.61 km) north of the highway linking Beirut and Damascus. One account of the mission states that the A-6E TRAM Intruders approached their targets at 20,000 feet (6096 m). If true, this would almost certainly be a mistake in an era when treetop level is the safest place for strike aircraft.

Into the Mission

Doppler Lange eased back on the stick, held throttle, and took 152915 upstairs from *Kennedy*. Lange tucked into formation, flying the Intruder smoothly and easily, perhaps apprehensive but certainly unaware that he was well into the final moments of his life.

Beside him, Bobby Goodman watched the Lebanese coast approaching. Goodman busied himself with bombardier-navigator tasks, paying close attention to his radar warning receivers (RWR). In his earphones were sparse messages between the various callsigns being used today, including CLOSE OUT (*Kennedy*'s E-2C Hawkeye), STEEL JAW (Mazach) and BROAD SWORD (the A-7E Corsair from *Independence* flown by that carrier's CAG, Commander Edward K (Honiak) Andrews). Radio discipline was good but it was obvious to Goodman that a fight lay ahead. *Indy*'s airmen were calling out SAMs (surface-to-air missiles) and other warnings. Referring to Andrews from *Independence*, not Mazach from *Kennedy*, someone shouted, 'Looks like CAG is hit at six o'clock!' Next, Goodman heard Andrews declare a Mayday—an emergency.

The other Intruder squadron aboard *Kennedy* VA-75—the unit which had introduced the Intruder to combat two decades earlier—went into a maze of criss-crossing Syrian missiles led by its skipper, Commander Jim Glover. A-7E pilot Andrews ejected near the Casino du Liban and parachuted into the sea off Beirut. As number three in his own flight, Doppler Lange kept 152915 boring relentlessly toward the target while crewmate Bobby Goodman looked out to see a missile trail that extended further into the sky than their own Intruder.

Lange's assigned target, it is understood, was an SA-9 radar site at Hammana. Bomb release was supposed to occur at 3,000 feet (4800 m) with the Intruder in a sharp dive. The Intruder was rushing toward its own shadow on a ridgeline above a village shrouded by an 800-ft (1400-m) peak where Syrian defence systems were sited. It is not known whether Lange ever saw the red-orange exhaust or the furling smoke trail of the Syrian missile that had his name on it.

Goodman, at least, never saw the missile that hit his Intruder. There was an enormous crunching sound. The aircraft was hit with a violent force and thrown into a nose-down attitude. 'Fireball!' someone in another plane shouted. 'You're hit! You're burning!' proclaimed another voice. Goodman saw the earth rushing up at him and knew he'd been hit, knew he should eject—but he would never remember actually doing so. Others saw 152915 tumbling in flames, The Intruder smashed dead-centre into a ridgeline, broke apart in a fiery spray of debris, and tumbled down the ridge.

Widely-published reports that Lange was still aboard the A-6E when it crashed are believed to be inaccurate. It is almost certain that Lange ejected at the very last instant. His parachute snapped open only 100 feet (30 m) off the ground, barely high enough to attain some 'lift' at the last instant. Lange may have had injuries, but not critical ones, when he struck the ground. Simultaneously, a piece of debris from the Intruder's canopy, or possibly his seat pan, came crashing down on him and violently severed his leg. With plenty of help apparently readily available in the form of numerous Syrian soldiers scrambling around Goodman and Lange, rushing to take them prisoner, Lange bled to death. It appears the Syrians put the deceased pilot back into the Intruder for reporters' photo sessions *after* they failed to make any attempt to help him.

Lange and Goodman had each checked out a pistol with ten rounds of ammunition as well as a survival 'beeper' radio for communication with rescue forces. There was no opportunity to use pistol or radio. Goodman was badly injured and rendered briefly unconscious by the ejection. He woke up to find himself being grabbed, manhandled, and rushed into a vehicle where Syrian troopers roughly stripped him of flight gear, wristwatch and wedding ring. *Kennedy* journalist-in-residence Wilson later analysed this cruel behavior by the Syrians and made it very clear that quick application of a tourniquet would have saved Lange's life. Newspapers (not Wilson's) played a cruel trick on wives and families by mistakenly publishing a caption about CAG Andrews (who was rescued) beneath a photo of Goodman (shown being seized by the Syrians). Some news stories were ambiguous about the fate of Lange (who was killed). Although he was identified incorrectly, the photo of a dazed, injured, blindfolded Goodman in hostile hands was widely published and seemed, in a sense, a symbol of American frustration and impotence.

And the air strike? The potent bombload aboard 152915 was still on board when the Syrians allowed newsmen to examine and photograph the Intruder's wreckage. CAG Mazach's Intruder suffered an equipment failure which made it impossible to drop its bombs. None of the eight other Intruders from *Kennedy* carried anything resembling a full bomb load. The accuracy of the eighteen Corsairs and Intruders from *Indy* has been questioned. At the time, first reports indicated that the air strike had been successful—the author of this work said so, too, in a companion volume about the A-7 Corsair—but revelations by writers Wilson, Bert Kinzey, Roger Chesneau and others make it exceedingly doubtful that the air strike was, in fact, a success or anything like a success. Wilson: '*Tuttle . . . was ordered to strike "at first light" in [a] communications breakdown . . .*'

Goodman was beaten, yelled at, tied up and taken to Damascus with Lange's body. He became a prisoner of war (POW) in a country which did not call him a prisoner, because of an action which no one called a war. Isolated, he was treated as a political prisoner for fully thirty days until Syrian president Hafez Assad—after ignoring Reagan administration demands for the return of the bombardier-navigator—released him into the hands of presidential candidate Jesse Jackson. Jackson made much of his role as a self-appointed envoy and, as much as he may have annoyed the people who practice diplomacy for a living, must be credited with accomplishing what he set forth to do.

Inevitably, controversy arose when a presidential aspirant who did not represent the administration practised amateur foreign policy of his own, intervening successfully with Assad to secure freedom for captive Goodman. Reporters noted that race was a factor in the Intruder crewman's release when Goodman, a black, was freed by Jackson, a black. When Bobby Goodman was safely reunited with his wife Terry on 4 January 1984, those same reporters found it necessary to inform readers that she was white. A man of rich humour, Goodman was remembered by squadron mates for his wit at VA-85's 'Intruder Ball' held before Lebanon at NAS Oceana, Virginia—when Hangar 122 was emptied of TRAM A-6E airplanes to make room for dance band and dancefloor. Although 'Black Falcons' is the nickname appended to VA-85's official history, 'Buckeye Blackeyes' is the name the men actually use. Far from being offended when flying buddies blackened their eyes with burnt cork for festivity at the ball, Goodman whitened his.

Within the Navy and without, Goodman was praised for his conduct in captivity. He resisted his captors, refused to cooperate with them, and—when others attempted to make a charade out of his release—he was particularly scrupulous. Whatever his feelings, he maintained a professional distance from, and avoided any display of solidarity with, or endorsement of, candidate Jackson. In short, Bobby Goodman conducted himself as a naval officer.

Despite his professionalism and his unwillingness to seek personal publicity, Goodman was unable to avoid becoming something of a celebrity as well as a hero. He and Terry were welcomed home by President Reagan—no fan of Jackson's—and appeared in the White House Rose Garden. He was exhaustively interviewed by Elisabeth Bumiller who writes Washington 'style' pieces about famous personalities.

Again the question: what about the air strike?

On balance it appears that the one-off mission in Lebanon on 4 December 1983 was less than an overwhelming triumph for American arms. As usual, the fact lay with civilians who kept changing their decisions about what they wanted—not with warriors whose sole job was to carry out orders.

Electronic War

Might the Lebanon air strike have fared even worse, had there been no Grumman EA-6B Prowlers available to jam and deceive the Syrians' defences? EA-6B crews such as Commander Holloway's spend years analysing air defence radars of every size and shape and maintaining the readiness to operate against them in a real combat situation. The EA-6B, as already noted, has been present at every crisis where naval aviation has played a role in recent years. Details are not known as to the role played by the EA-6B in the 4 December 1983 assault on Syrian targets in Lebanon, but it is fair to assume that some of the SAMs flying around the attacking A-6E and A-7 crewmen were inaccurate precisely because the Prowler was in the neighbourhood.

The effectiveness of the EA-6B Prowler in carrying out the electronic warfare role is, indeed, one of the 'good news' stories of the period. Air Force officers who fly the EF-111A Raven, which performs a similar job but in a somewhat different manner and with a crew of two instead of four, are said to have expressed their amazement at the capabilities of the EA-6B. It should be remembered that the EA-6B carries no weaponry of any kind, and yet is absolutely vital to any planned attack scenario.

Lessons Learned

If the US naval aviation community has one strength above all others, it is the flexibility to learn from mistakes. In the Lebanon case, it was not naval officers or naval aviators who made mistakes: the error lay not with the men in uniform but with the policy makers. Still, they *did* learn. From the moment the strike force began to recover aboard *Kennedy* and *Independence*, the learning process started. To send a 28-plane strike force and to suffer two aircraft shot down, one man rescued, one man captured, one man killed, was simply beyond the limits of what leaders like Sluf Tuttle were going to put up with. Although no public assessment of the Lebanon raid was ever

Commodore Gary F Wheatley, USN
(USN/E J Dail)

Rear Admiral Jerry O Tuttle, USN
(USN)

undertaken—Wilson strongly suggested that a congressional inquiry would be appropriate—it must be assumed that a rigorous 'in house' analysis was completed. Its conclusions have not been released.

In the long view, the 4 December 1983 mission cannot be rated a success, not when it failed to assist the intended American presence and influence in Lebanon—for, as it turned out, the Marines were soon withdrawn and the residual US presence, since then, had fallen prey to hostage-taking and hijacking. Scarcely a year later, terrorists in Beirut seized and later murdered CIA station chief William Buckley, one of several American hostages taken in that troubled capital. In 1985, the hijacking of Trans-World Airlines' flight 847, a Boeing 727, and the holding of its crew and passengers at Beirut forced world attention on US difficulty in influencing the Lebanese situation, with or without Intruders. Although the TWA prisoners were released following a brief period of public humiliation for the United States, the pattern of violence has continued unabated. In

1986, Anglican church envoy Terry Waite joined the many westerners held against their will in Lebanon.

For US naval aviation, the one-of-a-kind combat mission did, indeed, teach lessons. Never again would a sensible, solid plan be abandoned abruptly in favour of a course of action which seemed to reflect no planning at all. Never again would Intruder crews be rushed into a premature launch without proper ordnance. Never again would attack aircraft approach targets at 20,000 feet (6096 m). In the future, better use would have to be made of weapons intended for a modern-day defence environment, such as the AGM-88A HARM (high-speed anti-radiation missile) designed to home on SAM radars and the AGM-84A Harpoon air-to-surface missile, often used for anti-shipping work but effective against many targets in transforming the Intruder into a 'stand off' attack platform.

The next time Intruder crews and other naval aviators had to go into battle—in Libya—the story was very, very different.

Chapter 6
The Academe

Success stories often receive little attention. So it is with the Grumman TC-4C Academe, the US Navy's version of the Gulfstream I executive transport—not exactly a mighty A-6 Intruder attack craft, not by any means, but an essential part of the Intruder story. The TC-4C, which made its first flight in the form of aircraft 155722 at Calverton, New York, on 14 June 1967, piloted by D B Seaman and C E Alber, quietly passed a 20th anniversary while this volume was in preparation.

Grumman Aircraft Engineering Co began design in the mid-1950s, of a twin-turboprop executive aircraft intended for a front-cabin crew of two and 14 to 24 passengers. Powered by two 1,990-hp Rolls-Royce Dart Mk 529-8X engines, the prototype Grumman G-159 Gulfstream I (with civil registry N701G) flew for the first time on 14 August 1958 with none other than Robert K Smyth at the controls. In recent years, the familiar Gulfstream name has become associated with the Savannah, Georgia, firm Gulfstream Aerospace, which has Smyth as its chief executive—neither the company nor the pilot being, any longer, connected with Grumman. N701G, how-ever, continues to serve as an executive transport for its original manufacturer, regularly ferrying people between Washington, DC, National Airport and Grumman headquarters at Bethpage, New York. The author made the Washington–Bethpage jaunt aboard N701G on 13 January 1987 and found the air-craft to be an 'oldie but goodie'. In more than two decades, the only major change in appearance is a sign telling passengers that Grumman policy will no longer allow them to smoke while aboard.

Some 200 Gulfstream I executive craft have become ubiquitous at the world's airports. But it is the military trainer version which is now celebrating an anniversary, having achieved much whilst attracting little attention.

Based on the successful Grumman Gulfstream I executive transport, the elegant TC-4C Academe is equipped as a flying classroom to train bombardier-navigators for the Intruder
(Grumman)

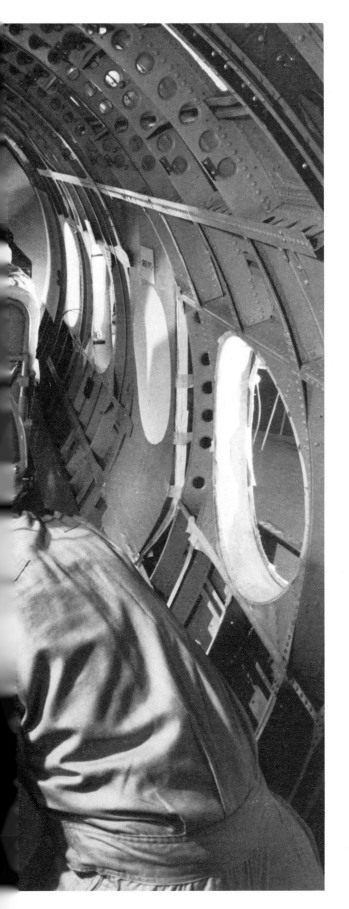

'Military Gulfstream'

The VC-4A was the US Coast Guard's variant of the Gulfstream I, two of which were delivered and used as VIP transports. The TC-4B was a navigation trainer/transport included in US Navy planning in the 1950s but never delivered. Finally, on 15 December 1966, eight years after civil N701G first spread its wings, the US Navy ordered nine G-159 Gulfstream I aircraft under the designation TC-4C. Scarcely noticed by anyone (except participants in a company-wide 'name the plane' contest), the 'naval Gulfstream' was officially given the nickname Academe. 155722 was officially accepted by the Navy on 15 March 1968. The nine TC-4Cs delivered to the Navy were assigned bureau numbers 155722/155730.

Equipped with the nose configuration of an A-6 Intruder, the TC-4C was, in fact, a flying classroom. Its interior was fitted with four identical replicas of the bombardier-navigator (BN) position in the Intruder, and the TC-4C became the standard trainer for several generations of BNs. Every right-seat crewman of an Intruder, from those who took the A-6A into combat in Vietnam in 1965 to those who used the A-6E TRAM Intruder to strike targets in Libya in 1986 . . . every single BN learned his job whilst sitting in the simulated Intruder crew position aboard the TC-4C Academe. Over the years, as the Intruder has been upgraded—for example, with addition of the TRAM system (target recognition and attack, multi-sensor)—the TC-4C has been correspondingly improved, as well.

TC-4C aircraft have served with the US Navy's two Intruder replacement air groups (RAGs), which in recent years have been rechristened fleet readiness squadrons (FRSs), these being the 'Green Pawns' of VA-42 at NAS Oceana, Virginia, and the 'Golden Intruders' of VA-128 at NAS Whidbey Island, Washington. TC-4C aircraft also served with the Marine Corps' Intruder training unit, VMAT(AW)-202, at MCAS Cherry Point, North Carolina, until 30 September 1986 when the squadron was deactivated and all Intruder training was consolidated into VA-128. Lt Col Randy Fridley, the final commander of VMAT(AW)-202, in addition to being an accomplished Intruder pilot with 231 combat missions in Southeast Asia, had also logged more than 1,000 hours in the TC-4C. During its existence, his squadron had produced 590 pilots and 611 bombardier-navigators while flying a total of 98,549 hours.

The interior of the TC-4C is equipped with identical replicas of the instruments and displays used by the bombardier-navigator (BN) in the cockpit of the Intruder (Grumman)

The TC-4C has kept pace with the advances introduced on successive Intruder models and features the TRAM (target recognition and attack, multi-sensor) turret fitted to the A-6E
(Grumman)

RIGHT
The success of the TC-4C as a training system is a matter of record, but the aircraft is also a hit with its pilots because it is so enjoyable to fly
(Grumman)

Successful Trainer

Considering that the statistics for the two Navy squadrons which employed the TC-4C are almost certainly even more impressive, the trainer variant of the Gulfstream I must be considered a success. In two decades of flying, only one of the nine TC-4C aircraft has been lost. It happened on 16 October 1975 when aeroplane 155723 belonging to VMAT(AW)-202 crashed in woodlands while attempting to land at Cherry Point. Sadly, all nine men on board were killed.

Commander Michael P Curphey, who flew the TC-4C with VA-128 during 1979–81, says that the Academe 'has almost no flaws'. Pilots note that it shares with the A-6 Intruder a tendency to be a little loose with nosewheel steering. 'But it's an absolute joy to fly', says Curphey. 'It requires some attention but it can be a lot of fun.'

The TC-4C has a maximum cruising speed of 355 mph under normal operating conditions. Ceiling is 38,000 feet, range 500 miles. The aircraft weighs 27,411 lb empty, 37,000 lb fully loaded. Wing span is 78 ft 4 in, fuselage length 81 feet, height 23 feet, and wing area 610.3 square feet.

It is not at all certain whether the US Navy will want to employ the familiar trainer when the new A-6F Intruder joins the fleet beginning in 1990. Under a programme initiated by recently-departed Secretary of the Navy John Lehman, it was found that Intruder bombardier-navigators did not need as much time in the TC-4C as had been previously believed. With 'sister ship' Gulfstream I executive transports still plying the world's airways after three decades, it has to be hoped that the successful life of the TC-4C Academe is not over yet.

Chapter 7
Intruder Variants

Although the Grumman A-6 Intruder and EA-6B Prowler have followed a rather straightforward pattern of development, any complete listing of all variants of the basic design—those actually built, and those merely proposed—requires a lengthy listing. In many cases, particularly with Grumman model numbers (which in some instances duplicate Grumman *concept* numbers assigned to entirely different designs), only exceedingly limited information is available—as we pore through the list of numbers assigned—the reader's indulgence must be begged. Much of what follows is far from literary in style and far from easy to follow. After considering various approaches, including the possibility of yet another lengthy index, the author decided that the following narrative was the best way to cover all Intruder designs. The full story, of course, is an on-going one and will not be finished in this century, but what follows is an attempt to sum up the principal Intruder variants:

A-6A (A2F-1)

The A-6A (originally A2F-1 and also the Grumman Model 128, with no suffix), as has been noted, first flew on 19 April 1960 at Calverton, the first aircraft being bureau number 147864 and also known to Grumman as 'Shop 1', the pilot being the irrepressible Robert K Smyth. The official acceptance date for the first airframe was also 19 April 1960. War clouds were rising and the A-6A was soon in combat, also as already noted in this volume, the 'A Model' being the first operational version in this aircraft type's long-running success story. One little-known aspect of the A-6A story is its role in the mining of Haiphong Harbour, the major escalation which launched the second American aerial campaign over North Vietnam.

A new war 'up North' had been brewing for months, witness the US decision to fly 'limited dura-

tion, protective reaction' air strikes—a decision which caused the 9 April 1972 shootdown of Intruder pilot Major Smith described in chapter four. At 8:10 am, on 8 May 1972 (the first day of the Linebacker campaign), President Nixon's decision to again send US warplanes to the Hanoi–Haiphong region—in effect, a cancellation of the bombing halt declared four years earlier by Lyndon Johnson—was translated into action when six A-7 Corsairs and three A-6A Intruders, the latter from Marine squadron VMA(AW)-224, launched from USS *Coral Sea* (CVA-43). The Corsairs went up north to swing around and head on a southerly course while the A-6A Intruders went straight toward the enemy homeland by flying due east. Commander Roger Sheets led the Intruders in at an altitude of 50 feet (17 m), the Intruder force carrying twelve 2,000-lb (908-kg) Mk 52-2 mines, these being very heavy and highly specialized weapons far superior to the Mk 36 destructor mines then in widespread use. So important was this mission to the outcome of the war, and so heavy were North Vietnam's defences, that the Joint Chiefs of Staff (JCS) had estimated that up to 30 per cent of the strike force might not return.

In fact, although missiles and MiGs threatened them, Commander Sheets' force succeeded in laying

The EA-6A model, in a remarkable 'before and after' sequence. Yeoman photographer Joe Handelman caught aircraft 151596 in early Naval Air Test Center markings at NAS Patuxent River on 21 April 1971, carrying a missile on the outboard pylon—a capability unique to this variant. A decade and a half later, 151596 appeared in front of Handelman's lens again, this time appearing at Andrews AFB, Maryland, on 28 December 1986 in ultra low-visibility paint with modex AF and side number 604 as a part of the 'Star Warriors' of VAQ-209. The bulbous fairing at the tip of the fin is a feature unique to the EA-6A model (Joseph G Handelman, DDS)

its mines at 9:00 am—the timing was important, as it coincided with Nixon's speech about the buildup in the fighting—and in returning to *Coral Sea* without suffering any losses. 'The mine is not a spectacular weapon but it is most effective', said Admiral Thomas Moorer, Nixon's JCS chairman. 'Just think what the United States could have saved if we would have mined Haiphong Harbour eight years sooner.' Today's Intruders have the capability to carry the aerial-laid, shallow-water Mk 50 series of mines as well as the Mk 60, an enCAPsulated TORpedo (CAPTOR) which is dropped with a parachute and can remain in place under the sea, active for up to a year at a time. The importance of mine warfare is forgotten by many—but not by Intruder crews.

In 1963 when the bulk of the A models were rolling off the production line, anticipated service life was 27 years. This measure of the airframe's durability has proven accurate and by 1990/1991, those airframes which began life as A-6As (even though subsequently converted to E models) will begin retiring at exactly the same rate as originally built, leaving the Fleet with 'new build' A-6Es and newer A-6Fs.

The production run of the 'A model' ended with aircraft 157029. Many A-6A Intruders, however, were later modified to A-6B, A-6C, KA-6D and A-6E standard. In all, including twelve machines which became EA-6A Electric Intruders (below), 488 aircraft of the A-6A version were built.

EA-6A

The EA-6A, which had been initially designated A2F-1H although it would have been properly designated A2F-1Q (Grumman Model 128D and 128H) and is also termed the **Electric Intruder**, made its first flight on 31 March 1963 at Calverton in the form of aircraft 147865, piloted by John Norris. Norris, incidentally, had been with Grumman since 15 August 1954, and is typical of the superbly qualified people in the firm's flight test stable. A list of

aircraft types flown by him in Grumman test programmes is a handy measure of his achievements: G-21, F9F-6, F9F-7, F9F-8, F9F-8T, F9F-9, F11F-1, F11F-1F, YAO (Mohawk), AO-1, AO-1C, AO-1D, OV-1D, TF-1, S2F-1, S2F-2, S2F-3, WF-1, A2F-1, A-6A, EA-6A, A3D-1, A3D-2Q, T-28B, SA-16 and UF-1. Now the manufacturer's Director for its Western Region, this pilot knows his stuff!

It was the Marine Corps, not the Navy, which first recognized the value of the Intruder as a possible electronic warfare (EW) platform. The Marines were in the market for a new machine, being well aware that the Douglas EF-10B Skynight, known as the Whale in the Corps, was too long in the teeth to continue in this role. As things turned out, the EF-10B Whale *did* have to soldier for a time longer, and went to war with the Marines in 1965. The EA-6A replaced the Whale in the combat zone the following year.

The intent was for the EA-6A to retain partial strike capability despite its principal mission as an electronic warfare support aircraft. Elements of the A-6A model's bombing/navigation system were deleted, however, to enable the EA-6A to detect, locate, classify, record and jam enemy radiation emissions. 147865, the second Intruder built and modified to become the first Electric Intruder, was essentially a testbed and, although flown in EA-6A configuration, was not fitted with the full suite of electronic systems associated with the Electric Intruder. It was soon joined by aircraft 148618, another modified A-6A which in the view of some deserves to be considered the EA-6A prototype; it flew on 16 April 1963. In all, twelve EA-6As were built by modifying A-6A airframes, these being followed by fifteen airframes manufactured as EA-6As from the outset, bringing to 27 the number built. The EA-6A variant is the only Intruder which is distinctively different in appearance from the other variants of the basic design, the difference being seen in the form of a bulbous fin-top fairing. The fairing was designed to

LEFT
KA-6D with friend. Until the advent of the F/A-18A Hornet, the Grumman A-6 Intruder and Vought A-7 Corsair were the principal attack aircraft of the US Fleet. 151821, modex NK, side number 521 drops the hose to feed fuel to a Corsair
(LTV)

ABOVE
The A-6A. The first version of the Intruder is seen in this rare study of Naval Air Test Center aircraft 149476 on a flight from Patuxent River, Maryland on 14 November 1962. This version of Patuxent's red-painted tail was unusual, and the paint scheme did not last long
(via David Ostrowski)

house various antennas and other electronic counter-measures (ECM) equipment. In addition, the fairing is understood to house the ALQ-55 comms jammer.

The EA-6A arrived in the Vietnam combat zone in November 1966 when it took up duties at Da Nang with squadron VMCJ-1, a Marine Corps 'composite reconnaissance' squadron which also operated the RF-4B Phantom. Subsequently, the EA-6A variant also served with the 'Playboys' of VMCJ-2 at MCAS Cherry Point, North Carolina, and VMCJ-3 at MCAS El Toro, California. Having Intruders and Phantoms in the same squadron at three location, corresponding to the three Marine aircraft wings, proved uneconomical and on 1 September 1975 this gave way to a new arrangement, with all of the reconnaissance Phantoms joining newly-formed VMFP-3 at El Toro while all of the electronic warfare Intruders went to newly-created VMAQ-2 at Cherry Point. In later years, the latter squadron operated the four-man EA-6B Prowler while a reserve unit, the 'Seahawks' of VMAQ-4, became the only squadron in the Corps with the EA-6A. More recently, the 'Star Warriors' of reserve squadron VAQ-209 and the 'Axe Men' of VAQ-309 began operating the EA-6A, the first time that this version appeared in Navy markings.

ABOVE
The A2F-1, as the A-6A variant of the Intruder was first called, is seen in this 17 May 1961 study at Calverton, which appears to be a view of the number one airplane, 147864, after exchanging its natural metal finish for Navy white and gray
(via David Ostrowski)

RIGHT
The EA-6A. Seen on 25 May 1966, aircraft 148618 was the Electric Intruder used for the variant's carrier qualification tests (carquals) aboard USS Kitty Hawk *(CVA-63) shortly after this simulated carrier launch taking place at NATC Patuxent River, Maryland*
(USN)

Grumman records indicate that at least two more company model numbers were assigned to this sub-type. Model 128D was a March 1960 study for a 'countermeasures version of the A2F-1H (EA-6A) airplane'. Model 128Q is listed as a June 1966 programme for 'EA-6A antenna flight evaluation'. As indicated, 27 EA-6A airplanes were built.

NA-6A

NA-6A is the designation applied to several A-6A Intruder airframes which have been modified for test purposes. One of these machines was the fourth Intruder built, aircraft 147857, rigged with large outboard electronic pods beneath the wings and a small, teardrop-shaped fairing on the left fuselage just beneath the cockpit. This machine is identified by one source as a test ship for the Iron Hand, or SAM suppression, role carried out eventually by the A-6B model, although it almost certainly served to test systems for night operations against vehicle convoys, the mission of the A-6C TRIM version (below).

TA-6A

The designation TA-6A, apparently an unofficial term, was used within the company to refer to the Model 128L, a proposed *three-place* version of the Intruder intended for the training role but never built.

EA-6B Prowler

The EA-6B, the well-known Prowler (initially the Grumman Model 128J), made its first flight on 25 May 1968 in the form of aircraft 149481, which had originally been A-6A airframe number fifteen, Don King being the pilot on the EA-6B maiden flight. Deliveries of production EA-6B airplanes began in January 1971 and the type was in combat in Southeast Asia by July 1972.

Not really a 'variant' of the Intruder at all, the Pro-wler is very much a wholly new aircraft, even though one reference source treats it simply as an Intruder with a 4 ft 6 in (1.37 m) extension to its nose plus a large fin pod like that found on the EA-6A but not on attack Intruders. Because of its major differences, not least being a four-man crew, the EA-6B is treated in a separate chapter. The EA-6B was still in production when this volume was being prepared, and seemed likely to be an important part of US Navy and Marine Corps airpower for the remainder of the century.

A-6B

The A-6B Intruder must have made a maiden flight at Calverton at some point—the first, that is, of the three major sub-variants encompassed by the A-6B designation—but Grumman has no record of when this occurred, who was pilot, or which airframe was involved. In fact, the original A-6B designation was for a clear-air attack version of the Intruder with no all-weather capability. It is unclear whether this was the same concept as a single-seat Intruder (Grumman Model 128G-12) proposed in August 1963 for the Navy's VAL (light attack) competition which led, instead, to the Vought A-7 Corsair, but in any event the 'daytime' A-6B airplane was never built and the designation went instead to an Intruder optimized for the Iron Hand, or SAM suppression mission.

At the outset of Southeast Asia fighting, the Navy was obliged to hastily modify a number of A-4 Skyhawks to carry the AGM-45A Shrike anti-radiation missile employed against North Vietnam's surface-

TOP
Close behind the A-6A design came the EA-6A, called the Electric Intruder and used initially only by the US Marine Corps. The EA-6A was operated in Vietnam and remains in service. Aircraft 156991, coded CY-19, of the 'Playboys' of VMCJ-2, the squadron which later became VMAQ-2, is seen during a visit to Kelly AFB, Texas on 13 November 1971
(Norman Taylor)

ABOVE
The NA-6A. This aircraft, actually the fourth Intruder built (bureau number 147867, was apparently a test ship for the later A-6C series. Distinctive features include a teardrop-shaped bulge just beneath the pilot and large, outboard wingtanks housing electronic gear
(Grumman)

TOP
*The NA-6A. Another view of the fourth A-6A Intruder
airframe (147867) serving in its role as an NA-6A testbed
for the eventual A-6C TRIM Intruder*
(Grumman)

ABOVE
*A very smart EA-6B Prowler from VAQ-129 'Vikings'
parked on the ramp at NAF Andrews on 24 February 1973*
(Joseph G Handelman, DDS)

to-air missile radars. The Air Force was better pre-pared with F-100F and EF-105G Wild Weasel air-craft for the same role. For a time, the A-6B seemed the solution though, in the end, it turned out not to be—Skyhawks and Corsairs carried out the bulk of the Iron Hand operations during the 1964–73 conflict.

The A-6B thus became something of a sidelight in the Intruder's history. Still, no fewer than three vari-ations of the A-6B model appeared.

The **A-6B Mod 0** was the initial configuration, lacking some of the sophisticated electronic fit found on later machines, but rigged for the AGM-78A Standard ARM anti-radiation missile. These appear to have been 'no frills' Iron Hand craft, produced with some haste to meet immediate requirements posed by the Southeast Asia fighting. Ten machines appeared in this series, all modified from A-6A stan-dard and carrying bureau numbers 149949, 149957, and 151588/151565.

The **A-6B PAT/ARM**, first flew on 26 August 1968 in the form of aircraft 155628 with Al Quinby and A J Beck on board, and was formally accepted by the Navy after what seems a lengthy delay on 13 June 1969. PAT/ARM is an unwieldy acronym for passive angle track/anti-radiation missile. The first half of that alphabet soup, the PAT system, was developed by Johns Hopkins Laboratories to

ABOVE
The original A-6B? The Grumman Model 128G-12 was a single-seat variant of the Intruder proposed in August 1963 for the Navy's VAL (light attack) competition which resulted, instead, in the Vought A-7 Corsair II. It is unclear whether the designation A-6B was intended for this aircraft, which exists today only in the form of a model displayed at Grumman's History Center
(Grumman)

TOP LEFT
The A-6B. Designed to attack North Vietnamese missile sites, the A-6B is seen here taking off on a stateside test flight carrying two AGM-78A Standard ARM missiles. Small nodules on the radome of the aircraft are antennas designed to detect enemy radar transmissions
(Grumman)

LEFT
A-6B Mod 1 TIAS (target identification and acquisition system) aircraft 151591 of the 'Blue Blasters' of VA-34, modex AB, side number 522, serving aboard USS John F Kennedy and seen at NAS Oceana, Virginia in June 1974. Barely visible to the naked eye are 'pimples' on the radome which characterize this model. Since Kennedy never served in Southeast Asia, this view is evident that the Iron Hand Intruder was deployed with the Atlantic Fleet
(Jerry Geer)

enhance the accuracy of Standard ARM (though, as with other Iron Hand aircraft, no one developed during the conflict a missile which could *continue* to home in on North Vietnam's radars after the enemy became aware of the threat and turned them *off*). Three machines of this type were converted from A-6A configuration and given bureau numbers 155628/155630.

The **A-6B Mod I** (or **TIAS**) first flew on 1 October 1968 in the form of airframe 151820 with C H Brown and R McDonnell as crew, and was accepted by the Navy on 30 April 1970. This version was an interim Iron Hand variant equipped with AN/APS-118 target identification and acquisition system (TIAS) developed by International Business Machines. The AGM-78 Standard ARM Mod/01 missile was employed. Six of these machines were converted from A-6A airframes and were assigned bureau numbers 149944/149955, 151591, 151820, and 152616/152617.

Apparently, the intent was that each squadron would have two A-6Bs co-mingled with its other Intruders. The 'Black Panthers' of squadron VA-35 apparently operated A-6A, A-6B and A-6C models all at the same time. According to Lieutenant Commander Tom Patterson, during a fall 1972 cruise by USS *Coral Sea* (CVA-43) which became the first 'post-war' cruise in Vietnamese waters, the 'Green Lizards' of VA-95 were equipped with three A-6B airframes in addition to about a dozen A-6A Intruders. According to Lt Cdr Roger Lerseth who flew the version, the A-6B was capable of operating alone against SAM sites since it had passive electronic countermeasures (ECM) equipment and could hunt out and destroy SAM sites without the help of other aircraft. The usual warload was two Shrike and two Standard ARM missiles with a fuel tank on the centreline station.

At one time, there were to have been at least several dozen A-6B airframes. Congressman Otis Pike, a former fighter pilot who could hardly be called impartial since Grumman was in his district, lit mightily into Secretary of Defence Robert S

TOP LEFT
The A-6C TRIM Intruder was developed to exploit the same technology used aboard the four Lockheed AP-2H Neptunes which were flown by squadron VAH-21 against the Ho Chi Minh Trail. The Navy had decided not to pursue the Neptune programme further. Eventually, it reached a similar decision with the A-6C
(Lockheed)

The A-6C. Aircraft 155667 was the first of the A-6C TRIM (trail-roads interdiction, multi-sensor), designed to attack the communist supply line known as the Ho Chi Minh Trail
(Grumman)

McNamara over what Pike called the 'sneaky' cancellation of a $22 million contract for fifty-four A-6B airplanes. In all, only nineteen A-6B Intruders were built.

A-6C

The A-6C (which may also have been known as the Model 129T, listed below) aircraft made its first flight in the form of airplane 151568 at Calverton on an unknown date with Al Quinby and Dennis Romano as crew, according to Grumman records. It would appear that the first airframe was not fully equipped with the systems intended for this sub-type, since Grumman history records indicate that the **A-6C Sea Trim** aircraft, the first being bureau number 155647, made its maiden flight on 11 June 1969 at Calverton flown by M J Burke and D R Cooke. The latter aircraft was accepted by the Navy on 26 May 1970.

The A-6C was an A-6A airframe equipped with electronic trail-roads interdiction, multi-sensor (TRIM) equipment to enhance its night and all-weather low altitude night attack capability, apparently with the North Vietnamese supply route, the Ho Chi Minh Trail, in mind. Previous Intruders had not been quite so specialized for operations against, for example, convoys of enemy trucks. The A-6C carried low-light level television (LLLTV) and the Black Crow receiver, a sensing device which detected emissions from vehicle exhausts.

In February 1969, it was reported that a $50 million contract had gone to Grumman for 'about fifteen' of these aircraft, noting that the TRIM sensor technology had originally involved both the Intruder and the Lockheed P-2 Neptune. In fact, four AP-2H Neptunes with the TRIM sensor conducted operations against the Ho Chi Minh Trail but the decision was taken not to pursue the Neptune programme further because, according to one report, 'the type was unsuitable for overland conditions'. Following early flights at the Grumman facility, forty-one flights of the A-6C TRIM aircraft were carried out at NATC Patuxent River, Maryland, ending in July 1970, to determine the sub-type's carrier suitability and maintainability and to determine the technical and tactical capabilities of the new sensors. It would appear that the A-6C was taken into combat thereafter by the 'Boomers' of VA-165. Twelve A-6C aircraft were converted from A-6A airframes, the eleven survivors eventually being converted to A-6E standard.

Perhaps the least known Intruder, the A-6C TRIM was designed to do the same job as the AP-2H Neptune in attacking North Vietnamese supply runs on the Ho Chi Minh Trail. Also pictured on the preceding page, 155667 was the test ship for the A-6C programme
(Grumman)

KA-6D

The KA-6D tanker version of the Intruder made its first flight at Calverton on 16 April 1970. The first aircraft was bureau number 151582, another former A-6A, and the crew for the first flight consisted of Chuck Sewell and D R Cooke. Official acceptance date for the first KA-6D was 15 May 1970.

In April 1966, the Navy demonstrated a possible Intruder tanker when the hose-and-reel assembly needed for the tanker role was fitted to the ventral equipment package (the 'birdcage') of A-6A bureau number 149937, employed to refuel an F-4B Phantom. 149937 which first flew as a tanker on 23 April 1966 was also employed for mid-air refuelling of another A-6A Intruder, aircraft 152584. At the time, a dedicated tanker was not a high priority item in the Navy, in part because several aircraft not dedicated to the purpose could function temporarily as tankers by employing 'buddy' systems. (The A-7 Corsair proved effective in the flight-refuelling role with a 'buddy' package, although the proposed KA-7F tanker was never built.) Eventually, with the Douglas KA-3B Skywarrior being phased out of inventory, the Navy 'came around' to its need for a flying gas station version of the Intruder and the KA-6D was not merely built but blooded. The 'Thunderbolts' of VA-176 took the KA-6D into combat in 1971. Thereafter, Intruder squadrons have usually consisted of ten attack and four tanker aircraft.

The KA-6D performs its tanker role very effectively but it transforms the bombardier-navigator into little more than a passenger, since he has no instruments or controls of any consequence on his side of the cockpit.

In an interview with Grumman officials, author

Thought to be a view of the 16 April 1970 first flight made by Chuck Sewell and D R Cooke, this view shows the first tanker airplane, 151582 (actually the 9th A-6A airframe) on takeoff from Calverton. The appearance of the designation on the Intruder's tail is unusual, and was not seen for long (Grumman)

RIGHT
KA-6D tanker with friends. In the late 1970s, a pair of F-4J Phantoms from short-lived squadron VF-191 fly near USS Coral Sea CV-43, accompanied over the Pacific by KA-6D Intruder number 152921 of the 'Green Lizards' of VA-95 (USN)

David Anderton found that the KA-6D satisfies Navy requirements until the year 2000, and that the Navy is not considering a KA-6D replacement. At the time of the interview, the Navy had a well-defined programme for conversion of about 45 to 48 A-6E airframes to KA-6D standard. All of these would be aircraft which had begun life as A-6As, had been converted to A-6E, but had not been retrofitted with the TRAM system (of which, more below).

Since so many Intruders have been converted from one mark to another, a few notes are appropriate about how these airframes became KA-6Ds. They received all new fuel tanks with two fuselage bulkheads being replaced (this was considered cost effective). There was extensive rework of the outer wing panels: changing the front and rear beams, rebuilding an outboard rib, all of these was tantamount to creating a new wing. There was selective overhaul of some components, such as actuators. The modification included a complete rewiring of the entire aircraft with installation of the latter day Omega global navigation system. And of course, the modification process involved removing the complete bomb aiming system leaving the KA-6D strictly a tanker. Of course, the flight refuelling system was installed in the ventral avionics bay known as the 'birdcage'. The KA-6D was designed to carry 10,000 to 15,000 lb of give away fuel, depending on the mission. It can transfer fuel at the rate of 300 gallons per minute.

In fact, fifty conversions of A-6A airframes to KA-6D standard, carried out by Grumman, were followed by sixteen more converted by the US Navy and a further four again converted by Grumman, bringing the total to 71 aircraft. Further conversions were expected to take place in the late 1980s.

A-6E

The A-6E Intruder (Grumman Model 128S) made its first flight at Calverton on 27 February 1970 when aircraft 155678, another of the ubiquitous A-6A air-

frame conversions, was taken aloft by M J Burke. The first *production* flight of an A-6E occurred on 26 September 1971 when aircraft 158041 was taken aloft at Calverton by G W Keyes and D B Marks. Official acceptance date for the latter was 1 December 1971.

Powered by two Pratt & Whitney J52-P-8B turbojets each rated at 9,300 lb (4128 kg) static thrust, the A-6E version employs a new central computer, multi-mode radar, and weapons release system. The AN/ASQ-133 digital computer and Norden AN/APQ-148 multi-mode radar were incorporated into the A-6E variant. One minor change: the AN/ALQ-100 deceptive jammer was located in the boom on A models but was moved to the wing root on the A-6E. If anyone wonders, there is a relief tube in the A-6E, attached to the side of the seat, with a hose; there is one for each crewman.

A bombardier-navigator who has flown in both says that the change from A-6A to A-6E is 'like moving from a Model T to a Cadillac'. The A-6E, at least in the later incarnations described below, is now the standard medium attack aircraft for the US Navy and Marine Corps. Before its final configuration was established, however, like Intruders before it, the A-

6E was found in sub-variants:

A-6E CAINS (carrier airborne inertial navigation system) aircraft employ the Litton AN/ASN-92 inertial navigation system also employed by the F-14A Tomcat and the S-3A Viking, rather than the AN/ASN-31 system of the other E Model Intruders. These airframes have built-in preflight checkout equipment and other items designed to ease the burden of maintaining an inherently sophisticated flying machine. A subtle change: these aircraft have an additional air-conditioning turbine to improve cooling and, hence, an additional air scoop. In 1987, with the TRAM Intruder (below) in widespread service, A-6E CAINS airplanes were scheduled to be no longer in service.

A-6E TRAM (the much-mentioned target recognition attack multi-sensor) came into existence when the first development flight on an A-6E TRAM Intruder was accomplished on 29 October 1974 at Calverton by F Wagner and D R Cooke in airplane 155673, without sensors installed. The first *production* flight of an A-6E TRAM Intruder occurred on 29 November 1978 when aircraft 160995 was taken aloft at Calverton by W E Bassett and D B Marks. Acceptance date of the latter was 14 December 1978.

Continuing through at least November 1976, exhaustive developmental flying was carried out in airplane 155673, which had a distinctive marking on the tail signifying the TRAM installation. This test airframe did not have the same internal 'fit' as the production A-6E TRAM Intruder, even after sensors were installed, but was essentially a proof-of-concept vehicle. Although its new capabilities greatly enhanced the TRAM Intruder's ability to detect, identify and attack targets, the new variant was different enough that it probably should have had a distinct designation. Minor glitches with hardware had to be ironed out at the same time that training methods and tactics were devised. One of the changes from earlier A-6Es was a different range of instruments for the bombardier-navigator.

The first squadron to operate the A-6E/TRAM Intruder was VA-165, deployed on board USS *Constellation* (CV-64) in 1977.

At the outset of the programme 32 A-6E aircraft were converted to the TRAM fit. A-6E TRAM aeroplanes are identified externally be a precision-stabilized sensor turret containing both infrared and laser equipment, and located beneath the nose. The turret is partly but not wholly retractable. The sensor

Latter-day KA-6D tanker. KA-6D bureau number 152927 of the 'Blue Blasters' of VA-34, modex AB, side number 521, was wearing an early version of the Navy's 'low visibility' paint scheme when photographer Don Linn trapped it at NAS Oceana, Virginia, on 11 October 1981. Until the advent of dull gray, KA-6D tankers were always distinguished by a stripe painted around the rear fuselage (Don Linn)

system provides the crew with real-time television imagery of non-visual or radar targets. The TRAM system integrates FLIR and laser sensors with the multi-mode target identification tracking and ranging in any light or weather conditions. TRAM makes it possible to view terrain features such as road patterns, cultivated areas, ploughed fields and wooded regions as well as traditional radar targets.

'In the A-6E TRAM you don't have to search for the target on FLIR, you go in on radar. You can save the laser [FLIR] which requires more attention to detail until you're closer in. The FLIR has a zoom on it, so that you can get a wide field view or you can shift and have a narrow field view. This is adjusted by a thumbwheel.' These are the words of a bombardier-navigator who may, some day, have to use A-6E TRAM to attack a bridge or an enemy air-field. 'The system helps you with everything ... depression angle [the angle between the path of the aircraft and the path of falling bombs], slant range [the distance to target when measured at that angle]. You get a better solution using the laser when you're closer to target.' The bombardier-navigator has a left-hand slew stick which can be stowed downward.

A-6E TRAM/DRS is the term for an upgrade of the TRAM Intruder with some minor changes in internal systems and the under-nose detection and ranging set (DRS), the first two production-equivalent DRS systems having been installed in two airframes in 1977 and having acquired 953 hours of ground and flight testing by 1979 when this addition to the basic TRAM Intruder was in production. Delivery of the DRS system began in February 1979 and after some delays began to reach the Fleet in 1980. In 1987, it was understood that all attack Intruders in the Fleet had the CAINS, TRAM and DRS installations. There remained, however, differences among A-6E airplanes which suggested that, probably, all should not have been bunched into one sub-variant.

As of May 1987, it was understood that 177 airframes in the A-6E series had been delivered by or ordered from Grumman as new manufacture while 228 Intruders of earlier marks had been brought up to A-6E TRAM configuration. Included in these figures were the six A-6Es to be built by Grumman in 1986 and 11 in 1987, after which the company was to shift to the A-6F in 1988. Visiting Patuxent on 27 February 1987, the author looked over A-6E TRAM Intruder 162182, which was at that time the newest Intruder in the Navy and which was to precede the first A-6F airframe on the production line, although additional A-6Es were expected thereafter.

A-6E Intruder before installation of TRAM. A five-ship formation from the 'Tigers' of VA-65 fly over a choppy Atlantic in the mid-1970s
(USN via David Ostrowski)

TOP LEFT
*The A-6E. Seen at NAS Oceana, Virginia, on 4 October
1982, this A-6E 160995, modex AG, side number 500,
belongs to the 'Tigers' of VA-65 embarked on USS
Eisenhower (CVN-69). TRAM turret has not yet been
installed*
(Don Linn)

LEFT
*A-6E TRAM Intruder. At NAS Oceana, Virginia, on 1
May 1982, with its somewhat colourful markings soon to be
replaced by a dull gray, 152599, coded AJ-512 of the 'Black
Panthers' of VA-35 has been participating in carrier trials
for USS Carl Vinson (CVN-70). The TRAM turret
appears just forward of the nose wheel*
(Don Linn)

ABOVE
*A-6E TRAM Intruder testbed. Aircraft 155673 was the
first Intruder equipped with the TRAM (target recognition
attack, multi-sensor) and did much of the early test flying for
the system*
(Grumman)

The A-6F was the next-generation Intruder, one of the highest-priority items on the US Navy's shopping list, intended to be the principal Fleet medium attack aircraft of the 1990s and encouraged strongly by John Lehman, who served as Secretary of the Navy until April 1987. The first full-scale development (FSD) A-6F Intruder was expected to be aircraft 162183 and was expected, while this volume was still in preparation, to be rolled out at Calverton in July 1987 and flown for the first time in late summer, 1987. Five airframes which began life as A-6E models, 162163, were to be FSD or 'pre-production' examples of the A-6F, each having some but not all of the features of the final A-6F which will reach the Fleet in the early 1990s. These first five machines, for example, were to be completed without the Boeing-built epoxy/composite wing which will represent a major departure from previous Intruder variants (as well as considerable cost saving) and will be employed on all operational Fleet A-6Fs.

Principal change, however, is not the new wing but replacement of the long-familiar J52 engines of all previous Intruder variants with non-afterburning 10,700-lb (4853-kg) thrust General Electric F404-GE-400D smokeless engines. Radar improvements will include sharper resolution, longer range and additional modes of operation, acquisition and tracking of tactical targets at more than twice the range of the E model. Inverse synthetic aperture radar (ISAR) processing will enable the bombardier-

Future Intruders will have smokeless engines to reduce their chances of being spotted by an enemy. This one does not (William R Curtsinger)

RIGHT
The A-6F. Grumman created this full-scale mockup of the A-6F Intruder at its Calverton, NY, facility prior to cutting metal on the first airframe in the new series (Grumman)

navigator to classify his targets at greater distances. The A-6F will also have an additional outboard wing weapons station to accommodate AIM-9L Sidewinder or AIM-120A AMRAAM (advanced medium-range air-to-air missile), to give the F Model Intruder enhanced capability to cope with any MiGs which might venture into its way. Externally, the re-engined, re-winged A-6F will differ from traditional Intruders primarily in having an additional dorsal scoop for cooling air.

Joseph Le Strange, 28-year veteran of Intruder development who witnessed the first A2F-1 flight in 1960 and is now working on the A-6F's internal systems, gave the author a look at the significantly different cockpit design of the A-6F. The cockpit is 'ergonometric', one of those buzz words which seems to mean crew-efficient. The crew will have five new multi-function displays rather than the three head-down cockpit displays or instruments in the E Model. The A-6F will have vertical and horizontal situation displays for the pilot and FLIR/radar and weapons management displays for the bombardier-navigator. It is remarkable to sit and watch how different functions of the attack mission can be shifted from one

of the five displays to another: because each display is identical in size and multi-function capabilities, each unit can be switched to another presentation in case of failure or change in mission requirements.

In addition, the pilot will be provided with a head-up display (HUD), replacing the current optical sight and improving visibility over the nose (although some erstwhile crews will continue the common practice of 'aiming' their aircraft by using the fixed in-flight refuelling probe as a 'gunsight'!). The HUD will provide a complete set of flight indicators, including air-to-air and air-to-ground symbology. For the first time, the bombardier-navigator will have, as the pilot has always had, a windshield 'wiper' ('so he can see out too', says Le Strange), though this is actually not a wiper blade but a blown-air system.

The A-6F will feature several improvements which are expected to reduce maintenance manhours per flight hour by 20 to 30 per cent. The aircraft will have an auxiliary power unit (APU) which eliminates the need for a power cart of other source to start the engines or to preflight-test onboard systems. A key maintenance improvement will come from an aircraft-mounted accessory drive (AMAD) on each engine. On earlier Intruders, removal of an engine requires disconnecting all accessories as well as fuel lines. With AMAD, accessories will be mounted on the airframe and driven by the power takeoff shaft, requiring only one disconnct in addition to the fuel lines to remove an engine. One potential complication: in the cockpit, the A-6F will be so different from previous models that crews may not be readily interchangeable. The Navy may decide to segregate A-6E and A-6F crews beginning with their initial Intruder training with the fleet readiness squadrons, with the result that men will not fly both variants.

Intruders That Weren't

At various junctures in this narrative, reference has been made to Intruder designs which got to the drawing board but were never built. Grumman 'Concept number' 128 airplanes beginning with Larry Mead's original series of M-winged designs have already been treated separately and some other 'non-Intruders', employing *Model* numbers in the 128 sequence, such as the single-seat proposal for the VAL competition, have already been mentioned. For the connoisseur of the arcane, the following is a wrap-up of Model 128 numbers assigned to Intruder projects:

Model 128A is listed in Grumman literature in one location as a 'general growth study' and in another as 'A-6 product improvement'. It seems to be a generic term to encompass early development from the original A2F-1 (A-6A) design.

Model 128B is identified in one document simply as 'AF-A2F-SR-195' and in another as 'Air Force Intruder—Dunmire'. The proposed Air Force attack Intruder apparently dates to October 1958, eighteen months before any Intruder flew, and represents the design skills of George Frederick (Fritz) Dunmire, one of the key people who laboured with Larry Mead and was present at the creation. The term SR-195 refers to an Air Force proposal, the final disposition of which the author could not learn.

The author is reminded of early days in Vietnam when an aircraft designed for the nuclear strike role, the F-105 Thunderchief, had only limited success attacking North Vietnamese bridges. Although the F-105 went on to glory, this happened only after much difficulty. It is tempting to speculate whether the early missions might have gone better had Dunmire's 'Air Force Intruder' reached squadron service and found itself available at the outset of hostilities.

Model 128C is the term for 'A2F-1 (A-6A) attack aircraft in Missileer airframe' and dates to March 1960, a month before the Intruder's maiden flight (see also Model 128E, below). This design is credited to 20-year veteran William Rathke, another key engineer and designer associated with Larry Mead's earlier efforts. Rathke, incidentally, went on to become LEM (lunar excursion module) engineering manager at Grumman, responsible for the Apollo landing vehicle which took man beyond the bonds of Earth during those exciting days of the late 1960s. (Today a remnant of *that* slice of Grumman history is the Astro Motel in Bethpage where the author overnighted whilst visiting Grumman. In better days, before anyone could imagine that a nation might go to the Moon and then lose interest, Astronauts partied in the motel whilst visiting for training sessions; today guests can still stay at the John Glenn room or the Gus Grissom suite.) Although this model was clearly an *attack* aircraft, the basic Missileer concept (of which, more below) was for a 'stand-off' interceptor to guard the Fleet. The actual Missileer interceptor based on the Intruder apparently encumbered the model numbers 128E and 128F (below).

Model 128D is the manufacturer's term dating to March 1960 for the EA-6A aircraft in its electronic countermeasures role (ECM) and is credited in company documents of Larry Mead.

Model 128E is the Rathke proposal for a Missileer version of the Intruder for the interception role and dates to June 1959. The proposal was classified CONFIDENTIAL at the time. The 128E version would have been powered by TF30 turbofans. A model of this unbuilt Intruder is on display at the firm's History Center.

Model 128F is an early Missileer concept, clearly predating the Model 128C and 128E (above) and other concepts stemming from the Navy's Missileer programme. Missileer, in 1959–61, seemed to be the way the Navy was going for carrier air defence. A slow, vulnerable airframe would go aloft, 'stand off' from approaching enemies, and launch long-range missiles to cut down an attack force. The AGM-54A Phoenix missile grew out of the Missileer pro-

The A-6F. Partially completed, the first full-scale development (FSD) A-6F airplane, bureau number 162183, being moved to the front of the Calverton assembly line in late 1986
(Grumman)

gramme, but no airplane did. Early on, it became clear that if the Navy ordered a Missileer it would be, not an Intruder variant, but the Douglas F6D-1 fighter, a never-built offspring of the F3D (F-10) Skyknight. The Missileer programme in its Douglas version was well-advanced when the Navy decided that another airplane, with utterly different characteristics, could defend its carriers—a then little-known type, the McDonnell F4H-1 Phantom. As has

been noted, the Navy decided instead on the Douglas F6D-1, and then went instead towards acquiring a Fleet defence force of F4H-1 Phantoms.

Model 128G is the company's term for the 'limited capability' (i.e., lacking night or all-weather capability), or VAL, design, which may or may not have been known as A-6B at one point. This is the single-seat design, already mentioned in its best-known Model 128G-12 form, intended for the VAL competition in which Grumman lost out to Vought's A-7 Corsair.

Model 128H, also mentioned previously, seems to be another design variation on the basic A2F-1H (EA-6A) Electric Intruder and dates to April 1962.

Model 128I is listed in some documents as PRAND and dates to October 1962. Nothing in the historical material currently available provided an explanation for this acronym.

Model 128J is another Electric Intruder designation, being listed in the company's records as 'EA-6A Tactical Weapons System'. It dates to March 1964. This model number is thought to have been applied to early proposals which in time resulted in the four-seat EA-6B Prowler.

Model 128K dating to December 1964, was an A-6A Intruder combat readiness improvement proposal.

Model 128L dating to June 1965, was a proposal for a *three-seat* version of the A-6 Intruder referred to within the company as the TA-6A, intended for the training role, and never built.

Model 128M is listed as 'P Band Jammer Pod' and refers to that electronic countermeasures (ECM) improvement, dated January 1965.

Model 128N was the nomenclature for a long-range navigation trainer version of the A-6A conceived in October 1965 for foreign export, but never built.

Model 128P of December 1965 was a Grumman submission for an unspecified attack system design competition and was never built.

Model 128Q of June 1966 was an EA-6A Electric Intruder antenna flight evaluation.

Model 128R was a RESCAP (rescue, combat air patrol) proposal for a version of the Intruder with tanker capability, guns, and visual navigation in lieu of many of the sophisticated instruments carried on attack Intruders. It dates to June 1966. At that time, the RESCAP or 'Sandy' function—escorting and supporting combat rescue airplanes and helicopters—was being carried out in Southeast Asia by the venerable Douglas A-1 Skyraider. Not until October 1972 was this role taken over by the Vought A-7D. Despite the need for a RESCAP or 'Sandy' aircraft, no machine intended specifically for this mission was ever built and this Intruder stayed on the drawing boards.

Model 128S also dated June 1966—apparently a busy month in the design shop—was an A-6A with improved electronics and with a proposed gun installation. No Intruder was ever built with an internal gun, although the idea was repeatedly considered.

Model 128T dated September 1966, is listed as the TRIM (trail-roads interdiction multi-sensor) version of the Intruder and appears to be synonymous with the A-6C.

Model 128U of October 1966, was a proposal for an ASW (anti-submarine warfare) version of the Intruder which never materialized. In 1966, the US Navy still had dedicated anti-submarine carriers such as USS *Intrepid* (CVS-11), and this never-built Intruder presumably would have been intended to

operate from such small carriers. It is interesting to speculate whether the Intruder might have become the US Navy's standard carrier-borne sub-hunter, instead of the Lockheed S-3A Viking which in due course received the job.

Model 128V was a flight evaluation proposal for Intruder tests with Doppler beam sharpening (DBS). It dates to November 1967.

Model 128W attributed to Grumman's J Cunniff and dated January 1973—a significant leap in time over the preceding model numbers—was a programme to evaluate EA-6B tactical support measures (TSM).

Model 128X which bears the name of Grumman's H Pabst and dates to March 1976, was an unspecified programme to evaluate the Intruder for foreign users. In fact, of course, no Intruder has ever been exported.

BOTTOM LEFT
Model 128E. This unbuilt Intruder was Grumman's proposal for the Navy competition for a Missileer, a fighter which could launch long-range missiles from a distance at aircraft attacking the carrier force

Model 128B. Yes, those are US Air Force markings, this design by Grumman's Fritz Dunmire having been intended to meet that service's SR-195 requirement in October 1958. The Intruder never served with the Air Force
(both Grumman)

Model 128Y is Grumman's term for an unspecified Intruder study attributed to the firm's J Alber and dated 25 April 1980.

Model 128Z dated 12 July 1983, was a study for enhanced surveillance capability by the Intruder and presumably refers to a proposed aircraft which would have had improved sensors.

Model 128AA of 30 August 1983 is the term which refers to a study of Intruder survivability and vulnerability.

Model 128AB the work of Grumman's R Scholly, refers to the EA-6B Prowler advanced capability (ADVCAP) aircraft and is dated 6 December 1983.

Model 128AC was apparently another version of Scholly's EA-6B Prowler ADVCAP work and also bears the 6 December 1983 date.

Model 128AC is Grumman's term for a Scholly-developed proposal dated 21 June 1984 for a replacement programme for encoding equipment on the EA-6B Prowler.

At this point in this long parade of alphabet soup, mercifully for the reader, Grumman's model number list moves on to the Model 129, a proposed ASW seaplane design bearing the name of engineer Mike Pelehach, and the long litany of proposed Intruders comes to an end, temporarily at least. The preceding list aought to give some impression as to the potential and diversity of the basic Intruder design.

Chapter 8
Combat: Prairie Fire

When it came time to test the mettle of A-6 Intruder crews against terrorist-related targets in Libya, American naval aviators were pitted against an adversary who had long challenged their country and countrymen. Operation El Dorado Canyon, the package of air strikes against Libyan targets which were unleashed just after midnight on 15 April 1986—itself preceded by Operation Prairie Fire on 24–25 March 1986—had been a long time in coming. The US had sought to act against terrorism in another country, Lebanon, with, at the very best, mixed results. In Libya, there was a determination to do it right.

The US had fought in Libya before. '. . . *to the shores of Tripoli*' is a key phrase in the US Marine Corps anthym. In 1801, 'Bloody' Yusuf Karamanli, a pasha who delighted in using his pirate fleet to seize US merchant vessels and hold their crews for ransom, incurred the anger of President Thomas Jefferson. Karamanli actually extracted 'protection money' from the US for a time in return for a promise *not* to seize prisoners. Jefferson had had it and, acting without any electronic masking or complex command-and-control procedures, sent the Navy and Marines to reduce the pasha's real estate holdings and put some of his vessels to the bottom. Peace was achieved the only way free men have ever achieved it, through willingness to employ force of arms.

Ever since the ascetic desert-bred Colonel Moamar Khadaffi seized power in Tripoli in 1969 and promptly began receiving heavy dosages of Soviet

Operation Prairie Fire. On 12 Ferbruary 1986, as Vice Admiral Frank Kelso's 6th Fleet force of three carrier battle groups prepares to confront Libya, flight deck crewmen aboard USS Saratoga *(CV-60) prepare Tomcats, Corsairs and Intruders for action. The actual fight came on 24–25 March 1986. Intruder in foreground carries 500-lb (227-kg) Mark 20 Rockeye bombs*
(USN)

military aid, oil-rich Libya had been unwilling to live by the rules of civility under which nations conduct themselves. Khadaffi had, however, been more than willing to train, finance, abet and support terrorists who killed Americans.

In the air, challenges have been persistent. On 21 March 1973, two Libyan fighters fired at a US RC-130 reconnaissance plane operating over international waters off their coast. In August 1981, after Khadaffi had announced a 'line of death' in international waters which he dared the 6th Fleet to cross, USS *Forrestal* (CV-59) with its F-4 Phantoms and USS *Nimitz* (CVN-68) with its F-14 Tomcats had found the Libyan air force challenging a legal right to operate at sea. Tomcat pilots found themselves repeatedly challenged by Libya's Su-22 *Fitter*, MiG-23 *Flogger* and MiG-25 *Foxbat* fighters, thought to be flown by pilots possessing more guts than skill. On 19 August 1981 after being fired upon first, two F-14 Tomcats of *Nimitz*'s 'Black Aces' of VF-41 engaged two Su-22s and handily shot them down. Not long thereafter, the author watched the Black Aces come off the boat for a visit to NAS Sigonella, Italy, successful in battle and hopeful that a message had been delivered to the leadership in Tripoli. But the message did not prevent two MiG-25 *Foxbat* fighters from confronting an EP-3 Orion over international waters on 6 June 1985, again displaying a hostility which transcended established international rules.

Enter the Intruder

Again, as had been done in Lebanon three years earlier, consideration was given to an air strike, or series of air strikes, which would be masked by the electronic wizardry of the EA-6B Prowler and carried out with the precision bombing capability of the A-6

Intruder. In 1986, the Intruder was more than a quarter-century old, still in outward appearance little different than it had been in 1960, and some might be forgiven if they forgot what a unique national asset it had become—the only carrier-based aircraft capable of operating in bad weather, at night, and delivering ordnance in a tricky and delicate situation where accuracy was everything and innocent casualties had to be avoided. Secretary of the Navy John Lehman was known to believe that the carrier air wing aboard each of the Navy's capital ships ought to have more Intruders embarked, even if it meant having fewer of some other aircraft type. Precision bombing could also be accomplished by the Air Force's F-111, but the latter required airfields, and airfields required permission from other sovereign governments to undertake the kind of action being contemplated against Khadaffi. Even then, not a single naval officer in the Intruder community was prepared to admit that the One-Eleven was nearly as good as the Intruder, not when the job really had to be done right. Many felt that had the *original* plan in Lebanon been followed (an all-Intruder strike at night), rather than the *ad hoc* plan which had to be

implemented at the last minute, the earlier experience in Lebanon would have been more fruitful. If Khadaffi seemed slow in learning lessons, the men in Intruders were not.

March 1986 Actions

It was not going to be a good idea for Khadaffi's airmen to become too bellicose. Embarked in the Mediterranean in early 1986 was carrier air wing thirteen (CVW-13) aboard the ageing but potent USS *Coral Sea* (CV-43), bringing the new F/A-18A Hornet into the region for the first time but also brandishing the A-6E TRAM Intruders of the 'Warhorses' of squadron VA-55.

A former Skyhawk unit which had fought in Southeast Asia and had been disestablished on 12 December 1975, the Warhorses had been re-established at NAS Oceana, Virginia, on 7 October 1983 and began to receive newly refitted A-6E TRAM Intruders on 25 January 1984. The squadron had a full complement of eighteen A-6E and KA-6D aeroplanes when it went to sea aboard *Coral Sea* on 2 October 1985. With action in Libya pending in March 1986, the Warhorses found themselves in a high state of readiness.

In March 1986, carrier air wing seventeen (CVW-17) was in the Mediterranean aboard USS *Saratoga* (CV-60) on a cruise which had begun the previous year. Vice Admiral Frank Kelso, commander of the

A KA-6D tanker from VA-55 'Warhorses' is towed along the flight deck during operations aboard the USS Coral Sea *(CV-43) on January 1986*
(USN/PH2 Rory Knepp)

Rear Admiral David E Jeremirah, Commander, Task Force 60, consults with another officer in the command and control centre of the USS Saratoga *during operations off the coast of Libya on 12 February 1986*
(USN)

6th Fleet in the Mediterranean. The previous October, with considerable help from EA-6B Prowlers, *Sara*'s Tomcats had carried out a successful middle-of-the-night interception of an Egyptian Boeing 737 carrying the hijackers of the cruise ship *Achille Lauro* and had forced the Boeing to land at Sigonella, resulting in apprehension of the hijackers. Following a brief foray into the Indian Ocean via the Suez Canal, *Sara* was back in the Med in March. The carrier's medium attack squadron was the 'Black Falcons' of VA-85 commanded by Commander Bob

Day. (The names of aircrews who flew against Libya are still being withheld and will not appear in the narrative which follows.)

VA-85's A-6E TRAM Intruders were equipped to carry the AGM-84A Harpoon anti-shipping missile, a relatively new asset in the ordnance inventory which was not to remain unblooded for long. Developed by McDonnell Douglas Astronautics and first fired from a P-3B Orion in May 1972, the Harpoon had been through a relatively trouble-free development programme and could now be carried by naval aircraft ranging from the Orion to the A-7E Corsair and A-6E TRAM Intruder. 12 ft 7 in (3.84 m) in length, with a body diameter of 13.5 in (3.43 m) and a launch weight of 1,160 lb (526 kg), Harpoon could range out at a speed of Mach 0.75 to strike targets 57 miles (92 km) away, having a 500-lb (227-kg) penetration/blast warhead with impact/delay and

*An A-6E Intruder from VA-55 comes to a stop after
recovering aboard* Coral Sea *on 29 January 1986*
(USN/PHAN Salman)

proximity fusing. It could slice through the metal skin of a warship and then explode, wreaking havoc. It provided Intruder crews with the 'stand off' capability to inflict harm whilst remaining outside an enemy's defences.

USS *Coral Sea* (CV-43), of which more will be heard later, was backing up *Saratoga*. *Coral Sea*, as has been noted, boasted carrier air wing thirteen including the 'Warhorses' of VA-55 under Cdr Rob Weber. The carrier task force—following a 14 March 1986 decision by President Reagan clarified by a meeting three days later at the American Embassy in London between Defense Secretary Caspar Weinberger and Admiral Kelso—was ordered to challenge Khadaffi's territorial claims by showing the flag in international waters off his coast.

The Weinberger–Kelso meeting had laid the ground work for yet another effort to deliver a message to Colonel Khadaffi.

Operation Prairie Fire

For more than a month, Libyan fighters had been following their pesky habit of scrambling to intercept Tomcats from the 6th Fleet carriers. Weinberger and Kelso ironed out the details for Operation Prairie Fire, the March 1986 effort against Libya which began as an innocent exercise in maritime law—a demonstration by the US of its right to sail in international waters. Kelso was to go into harm's way by challenging Khadaffi head-on, but only after strengthening his fleet with a third carrier battle group. USS *America* (CV-66) was en route to join *Sara* and *Coral Sea*. *America*'s contingent included the 'Blue Blasters' of VA-34, headed by Commander Richard Coleman who had flown 147 combat missions in Vietnam and accomplished no fewer than 800 carrier landings. VA-34's Intruders were also equipped to handle the Harpoon missile.

Admiral Kelso returned to his flagship USS *Coronado* (LPD-11) and issued orders to his three carrier battle groups as well as the missile cruiser USS *Yorktown* (CG-48) which was instructed to challenge Colonel Khadaffi's territorial claims by crossing his vaunted 'line of death' into the Gulf of Sidra—recognized by all nations except Libya as international waters.

At 1:52 pm, 24 March, *Yorktown* detected Libyan preparations at the launch sites for SA-5 *Gammon* surface-to-air missiles. Moments later, with *Yorktown* tracking the event, two SA-5s were fired at a brace of Tomcats 70 miles (110 km) off the Libyan coast. *Yorktown* employed its own electronics to foil the missiles. An hour later, two MiG-25s flew out towards the Tomcats but declined to engage. As evening approached, at 6:45 pm, more SA-5 missiles were launched at US fighters. Admiral Kelso had committed his three carrier battle groups under rules of engagement which required nothing further for him to retaliate.

TOP
A-6E TRAM, side number 510 of VA-55, lands on Coral
Sea *during operations in the Mediterranean in January 1986*
(USN/PHAN Salman)

ABOVE
*Flight deck crewman work on a Mark 7 blast defector aboard
the USS* America *(CV-66) as an MD-3A tow tractor pulls
a KA-6D tanker along the deck at the start of the day's
flying*
(USN)

TOP
*A Prowler pilot prepares for a mission during flight
operations aboard* Saratoga *on 28 January 1986*
(USN/PH1 William A Shayka)

ABOVE
A starboard quarter view of an Italian-built Assad-*class
missile corvette of the Libyan Navy underway on 8 January
1982. This vessel was not actually attacked by A-6s*
(USN)

'Birds of a feather, flock together'. A-6E TRAMs and KA-6Ds from VA-85 'Black Falcons' at NAS Oceana on 24 April 1986, having just returned from engagements with Libyan missile boats in the Med aboard Saratoga
(Joseph G Handelman, DDS)

LEFT
The CAG's A-6E TRAM, bureau number 151573, modex AA-500, from CVW-13 aboard Saratoga displays a missile boat kill downstream of the wing root
(Joseph G Handelman, DDS)

TOP
A general view of the CAG's A-6E TRAM parked at NAS Oceana on 19 April 1986
(Joseph G Handelman, DDS)

ABOVE
A-6E TRAM, bureau number 161662, modex AA-503, of VA-85 also scored a missile boat kill off the Libyan coast
(Joseph G Handelman, DDS)

TOP
A close-up of the missile boat kill marking applied to 161662,
illustrated on the preceding page
(Joseph G Handelman, DDS)

ABOVE
A 'Blue Blasters' A-6E TRAM, the CAGs machine from
America, *displays a resistance to totally toned-down*
markings. Serial 161231, modex AB-500
(Joseph G Handelman, DDS)

TOP
Detail of VA-34 tail marking worn by 161231
(Joseph G Handelman, DDS)

ABOVE
KA-6D tanker, bureau number 155604, modex AB-524, of
VA-34 'Blue Blasters' parked at NAS Oceana on 13
September 1986
(Joseph G Handelman, DDS)

TOP LEFT
Detail of MiG-23 Flogger *and missile boat kill markings painted on A-6E, bureau number 161681, modex AK-502, of VA-55 'Warhorses'*

LEFT
A general view of 161681 parked at NAS Oceana on 13 September 1986. VA-55 were embarked on Coral Sea *when they went into action against Libyan targets*
(both Joseph G Handelman, DDS)

ABOVE
A Libyan Navy Nanuchka II *class missile corvette burns after being hit by an AGM-84 Harpoon missile fired by an Intruder from VA-85 'Black Falcons' on 25 March 1986. The ship was later reported as sunk*
(USN)

Kelso's force now unleashed a series of retaliatory strikes with a mixed force of Corsairs, Intruders and Hornets, while EA-6B Prowlers scrambled Libya's electronic defences. The 'Black Ravens' of VAQ-135 under Cdr Denny Major had the distinction of being the first Prowler squadron to operate from the deck of *Coral Sea* and, as one observer put it, they 'jammed the hell out of Khadaffi'. More on how the Prowlers reached the scene, and thoroughly confused airplane markings enthusiasts in the process, will soon be detailed.

Khadaffi's forces responded by committing their navy in an attempt to sink US warships in the Gulf of Sidra. Apparently at 8:30 pm, four A-6E TRAM Intruders, two each from VA-34 on *America* and VA-85 on *Sara*, went against the closest patrol boat, a 160-foot (56-m), 260-ton *La Combattante II*. The VA-34 Intruders became the first to employ the Harpoon missile in combat, and pulverized the corvette but did not sink it. VA-85's Intruders administered the *coup de grace*, with Mk 20 500-lb (227-kg) Rockeye bombs. The ability of the Intruder to carry out such missions at night—or at least, in deceptive and fading twilight which amounted to night—was proven once again.

Intruder versus Vessel
Many of the Libyan combatants were 780-ton *Nanuchka II* class missile boats racing at 25 knots toward the US naval force. This vessel carried four SS-N-2C *Styx* medium-range missiles and an SA-N-4 surface-to-air missile system as well as 57-mm guns. At 9:06 pm, more SA-5s were launched and warplanes from Kelso's fleet responded by attacking a missile site. VA-34 and VA-85 continued to engage Libyan boats as the night continued, proving not merely the Intruder but the stand-off qualities of the Harpoon missile as well.

Had the corvette not gone to the bottom of the sea—its demise hastened by the fact that an E-2C Hawkeye was tracking it from the moment it left port—the vessel might well have threatened Kelso's fleet with sea-skimming, 460-lb (224-kg) *Otomat* missiles with a range of more than 50 miles (80 km). Now, VA-85's Intruders pounded a *Nanuchka II* to prevent the missile boat from posing a similar threat with its Styx missiles.

The exact sequence of the Prairie Fire missions is still not completely clear. A history published by VA-85 states that 'Libyan patrol boats, armed with surface-to-surface missiles, put to sea at night [24/25 March] and headed toward [*Saratoga*'s] battle group. A-6 Intruders attacked the patrol boats before they could reach their firing range. VA-85 conducted five of the eight naval air attacks against the Libyan patrol boats. All five scored direct hits.' A *Nanuchka II* limped back to port in Benghazi at 10:30 pm after being pounded by Rockeyes from VA-85's Intruders, and is listed as having been damaged.

Also to the east on the active night of 24–25 March, Libyan patrol boats and corvettes near Benghazi attracted the attention of Kelso's naval force. VA-55 and VA-85 engaged. At fifteen minutes past midnight on 25 March, in a rare surface-to-surface engagement, *Yorktown* attacked and sank a Libyan missile boat. A half-hour later, aircraft—almost certainly Intruders, but which squadron is not clear—again attacked an SA-5 missile site. On 25 March 1986, Tuesday, two more attacks were carried out on SA-5-related radar sites.

Another attack on a *Nanuchka II* was carried out by VA-85 Intruders apparently in clear daylight on 25 March, working in close coordination with an E-2C Hawkeye. A 'Warhorse' Intruder of VA-55 from *Coral Sea* had dropped Rockeyes on the missile boat and slowed it down, but the vessel was still boring toward the US armada. VA-85 thus mounted a Harpoon strike. With the Intruder's bombardier-navigator feeding information about the vessel's whereabouts into the Harpoon's guidance seeker, the crew launched the missile and saw it detonate, setting the boat afire. The vessel went down. Another Intruder pilot was treated to the sight of its 60-man crew leaping into the sea.

Prowler Role

The presence of the EA-6B Prowler during the Prairie Fire campaign was in part the result of some fast shuffling and bolstering of assets. As has been mentioned, *Coral Sea* did not originally possess a Prowler contingent. At the close of 1985, the afore-mentioned 'Black Ravens' of VAQ-135 were still assigned to *America* which was then scheduled for workups in early January. Due to the climate of increased tension in the Mediterranean, the squadron was tasked at that point to join *Coral Sea*. On 1 January 1986, five Prowlers (plus a spare which turned

ABOVE
The result of a Harpoon missile attack by a VA-85 Intruder against a Libyan corvette, The ship is dead in the water and beginning to settle by the stern
(USN)

RIGHT
Two views of the double-scoring A-6E TRAM, bureau number 159317, modex AK-501, of VA-55 'Warhorses' parked at NAS Oceana on 13 September 1986. In common with 161681 pictured on page 156, this Intruder destroyed a MiG-23 Flogger *and a missile boat*
(Joseph G Handelman, DDS)

back after tanking) passed through NAS Oceana, Virginia, en route to Rota, Spain. The men of the squadron actually preceded the Prowlers themselves, which were flown by spare crews, and picked up the aeroplanes as they arrived at Rota. From there, they flew to Sigonella, and had three aircraft aboard *Coral Sea* by 3 January. Throughout the following months the 'Ravens' operated from both *Coral Sea* and Sigonella, having a maximum of three airframes on board *Coral Sea* at any given time, the vessel being too crowded to handle more. Due to the rushed

LEFT AND BELOW LEFT
Hastily reshuffling in the EA-6B Prowler community led to some confusing markings and tailcodes when the 6th Fleet went into action against Libya. Although aircraft 160788, coded AA-607, still has 'Saratoga' painted on its fuselage (only barely visible to the eye), VMAQ-2 was, in fact, hastily put aboard USS America *(CV-66) for the Libyan operations after its own Prowler squadron, VAQ-135, was usurped by USS* Coral Sea *(CV-43). By April 1986, when Prairie Fire was over and Khadaffi was still supporting terrorism through the bombing of a nightclub in Berlin, aircraft 159909, coded AB-605, was in correct markings for its appearance aboard* America *(upper photo)*
(USN)

BELOW
EA-6B Prowler, bureau number 159586, modex NJ-903, of VAQ-129 'Vikings' rests at NAF Washington, DC on 29 November 1986
(Joseph G Handelman, DDS)

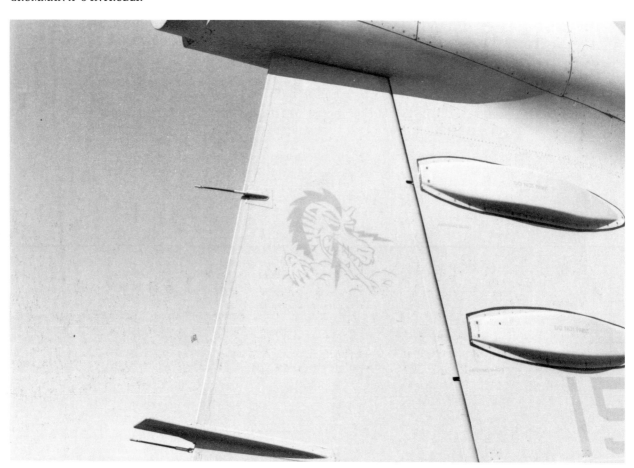

TOP LEFT
Close-up view of the low-visibility artwork on the rudder of an EA-6B (156481) belonging to VAQ-130 'Zappers'
(Joseph G Handelman, DDS)

LEFT
A general view of the same aircraft parked at NAF Washington, DC, on 18 October 1986. This Prowler is assigned to the Air Wing on Kitty Hawk
(Joseph G Handelman, DDS)

ABOVE
A weathered EA-6B Prowler, bureau number 158035, modex NJ-914, of VAQ-129 'Vikings' on the ramp at NAF Washington, DC on 30 August 1986
(Joseph G Handelman, DDS)

nature of their departure, the 'Ravens' took whatever airframes were available, creating a mishmash of east- and west-coast modexes (tailcodes) which has confused spotters ever since. In due course, the Prowlers were painted with proper 'Raven' markings and in the meanwhile, VAQ-135 logos were painted beneath the aircraft side numbers—the three-digit numerals on the nose which designate an airframe's place within a carrier air wing. Originally painted with side numbers commencing with 600 for their intended *America* commitment, the EA-6Bs received side numbers commencing with 620 whilst on *Coral Sea*.

Nobody sacrificed more than the VAQ-135 people. Due to space limitations on *Coral Sea*, the junior officers were quartered in a converted spud locker.

When *America* was deprived of her usual EA-6B Prowler squadron (the 'Ravens' having been moved before the decision to add *America* to the Prairie Fire armada), she deployed on the final phase of pre-cruise workups without EA-6Bs. At that point, the Marine Corps' 'Playboys' of VMAQ-2 were tapped to join *America*. 'Q-2' began selecting aircraft and crews, there being no shortage of volunteers 'now that', as one aviator says it, 'everybody knew a storm was brewing'. *America* eventually went into the fray with VMAQ-2 detachment wearing an AB modex. And if the Prowlers did a superb job with Prairie Fire in March, they were to become indispensable a few weeks later.

Prairie Fire was a huge and complex operation and it was, beyond question, successful. Certainly, it provided evidence that the Intruder is still in the prime of life. Isolated actions continued during daylight on 25 March and the *International Herald Tribune* headlined the situation: '*US Keeps Up Attack . . . Patrol Boats Are Sunk.*' Pentagon spokesman Robert B Sims did his duty and enunciated the obvious: 'We have been given ample evidence of hostile Libyan intent both by missiles and surface ship movements in the past two days, and we are going to protect ourselves.' From his grave, pasha Karamanli could have been excused for sighing with relief that he'd never come up against VA-34, VA-55, VA-85, or the hardworking EA-6B Prowler squadrons which were the *sine qua non* of the prolonged engagement.

'*Battle in the Sea of Blood*' headlined the British tabloid *Today*. According to Karen DeYoung of the *Washington Post*, who managed to get out a thousand words on Prairie Fire without once mentioning the EA-6B Prowler, US naval aviators had found the Soviet-built SA-5 missile to be a formidable opponent. DeYoung correctly quoted Admiral Kelso as saying, 'It's a very fast and capable missile', but seemed unaware that Corsairs and Hornets had employed the AGM-88A HARM (high-speed antiradiation missile) to cancel out any threat which had not already been confused and disrupted by the Prowlers. Members of the Prowler community will now

Everything down and out, a KA-6D tanker approaches the
deck of Coral Sea *after an in-flight refuelling mission over*
the Mediterranean on 29 January 1986
(USN/PH2 R Knepp)

LEFT
EA-6B Prowler, bureau number 156481, modex NG-606, of
VAQ-130 'Zappers' shares the ramp with CT-39G 160053
at NAF Washington, DC. The CT-39G features the longer
fuselage of the Sabreliner 60 business jet
(Joseph G Handelman, DDS)

receive an apology, at this juncture in the narrative, because their contribution is not celebrated in greater detail: their work is, by its nature, not amenable to publicity. VAQ-135, VMAQ-2 and the other Prowler detachments committed to Prairie Fire deserve as much credit as anyone for the success of this latest effort to get a clear message into Colonel Khadaffi's tent.

Richard Halloran of the *New York Times*—let it be said, a journalist of very great distinction, but not one who regularly reports on air warfare—noticed that 'new jet tactics' had been employed against Libya. ('Pilots kept well away from targets'.) Halloran:

'The tactics were developed after criticism of the performance of navy fighter planes in Lebanon in 1983. The criticism came from within and outside naval aviation, including the navy secretary, John F Lehman.' To a degree, Halloran was stating the obvious but he, rightly indeed, knew about Prowlers:

'The Navy flew Prowler aircraft packed with electronics to determine, or 'steal' Libyan radar frequencies so US missiles could ride the beam down to the Libyan radar and destroy it . . . In addition, the Prowlers jammed Libyan radar scopes so operators could not locate US aircraft, and they scrambled Libyan communications. They emitted false radar signals to deceive Libyan ground controllers, making them shoot away from Navy planes.' A *Times* colleague, Edward Schumacher, noted that the US air strikes had forced scores of Russian military advisors in Sirte to huddle for cover inside a local airbase to avoid

ABOVE
Groundcrew refuelling an A-6E of VA-128 'Golden Intruders' at NAS Fallon, Nevada, in December 1977 (USN/PH1 Randy Emmons)

TOP RIGHT
The crew of an A-6E keep a sharp lookout as their aircraft manoeuvres aboard Coral Sea *in April 1986. Although detachable, the refuelling probe is invariably fitted* (USN)

RIGHT
Another successful trap for a VA-55 Intruder aboard Coral Sea *as the arrestor cable begins to slacken and a flight deck crewman beckons the aircraft to fold its wings prior to parking on 29 January 1986* (USN/PH2 Rory Knepp)

OVERLEAF
An MD-3A two tractor manoeuvres an A-6E on the flight deck of Saratoga *in front of a tightly packed bunch of Tomcats and Corsairs during operations off the Libyan coast on 12 February 1986* (USN)

A crewman aboard the Coral Sea *takes five on the wing of an Intruder parked on the flight deck* (USN)

being drawn into the conflict with Kelso's 6th Fleet.

Not a single Intruder, or other naval aircraft, was scratched, dented, or even touched during the Prairie Fire effort. Colonel Khadaffi had been dealt a blow which even a humble desert nomad should have been able to understand.

Khadaffi had, it seems almost certain, sponsored the 27 December 1985 terrorist attacks on innocent civilians at the Rome and Vienna airports. But critics said that the proof of his role was lacking. One long-time US government expert on Libya, whilst acknowledging Khadaffi's role in some terrorist activities, opined that this self-styled modest man of the desert was being given a bum rap. Khadaffi, after all, lived in a tent, prayed on a rug, sought solace in the desert, and foreswore luxuries which could have been available to him had he lusted for them. The expert failed to mention *Nanuchka II* missile boats, MiG-25 fighters and SA-5 missiles—as well as the debris bunched inside pools of blood at Rome and Vienna where the innocent had been slaughtered. The criticism was, none the less, important: even the US acknowledged that direct, conclusive evidence of the terrorist connection was hard to come by.

Until West Berlin, Khadaffi *was* supporting terrorist attacks and some of them were being carried out against US forces. Although the evidence may have

seemed less than persuasive to critics who were able to find other culprits at Rome and Vienna, West Berlin provided the 'smoking pistol' (undeniable evidence) that had heretofore been lacking. The US obtained clear proof which linked Khadaffi to the 4 April 1986 bombing of La Belle discotheque in West Berlin, a gruesome act coming *after* Prairie Fire which killed one US soldier and a Turkish woman and wounded 250 others, including 50 American servicemen enjoying a weekend respite from duty. According to one published report, the US National Security Agency intercepted Libyan communications confirming that the disco bombing was merely a kind of trial run—that Khadaffi was extolling a whole series of anti-American attacks around the world, including at least ten more in West Berlin.

The Grumman A-6 Intruder had been used effectively and well, and there was no doubt that it would continue to be a vital part of the fleet. Intruder aircrews, some of whom had watched Libyan SAM sites go up in smoke or observed Libyan sailors leaping into the sea, were fully justified in feeling that they had done a superb job. Under a variety of conditions, with different ordnance, the Intruder had dealt a blow to undeniably military targets without inflicting unnecessary civilian casualties. The present-day version of the Intruder, with its TRAM system and Harpoon capability, was a tough act to beat. Incredibly, however—perhaps because the La Belle discotheque attack had been long in planning and could not be cancelled at the last minute—the US had gotten the goods on Tripoli's mercurial leader not before Prairie Fire, but after. Incredibly, Prairie Fire had not, in itself, caused an end to the Libyan colonel's adventurism.

The message, it seemed, would have to be delivered in a more forceful way.

Crewmen scrub clean the area around one of America's *catapults in readiness for a busy day of flight ops in April 1986. A Sidewinder-armed Tomcat of VF-102 'Diamondbacks', a Tomcat of VF-33 'Tarsiers' and an EA-3B Skywarrior are parked behind the Intruder in the foreground*
(USN)

Chapter 9
Future Intrusions

Intruder crews never stop learning. The new pilot or bombardier-navigator meets the A-6 Intruder for the first time at the RAG—for the purist, the fleet readiness squadron (FRS), meaning the 'Green Pawns' of VA-42 at Oceana and the 'Golden Intruders' of VA-128 at Whidbey. But even the most senior A-6 crewman, who may be a CAG (carrier air wing commander) or the skipper of a squadron, will find himself going back to school again and again. Some of the classes are carried out at the location previously described as one of the loneliest in the world—NAS Fallon, Nevada, 60 miles (91 km) from Reno. Not as isolated as it was in earlier years but still the location of some of the Navy's best target and range facilities, Fallon is headquarters for 'Strike U' (for university), which is more formally known as the Naval Strike Warfare Center. Movie-goers familiar with Maverick and Goose will know of a place called Top Gun which hones the Navy's best fighter pilots. Strike U, in the same manner, sharpens the talents of the Navy's best strike air crews. Class is held in the great open arena of the sky, and learning is accomplished with live ordnance. It isn't book learning, but there *is* a book, called *Rules to Live By*, which tells the strike crew member how to fight, to survive, and to win. One rule: 'Use surprise, deception and fight dirty!'

Strike U was established on 15 September 1984 at the personal order of that well-known Reservist bombardier-navigator, Secretary of the Navy John Lehman. Its first skipper was Captain Joseph Prueher, who was followed in due course by Captain Jay Finney. At the beginning, the Naval Strike Warfare Center's purpose was to train Corsair, Intruder and Hornet crews in the strike mission in much the same way that Top Gun produces Tomcat experts. In due course, it became clear that truly realistic strike training could be accomplished only during full-scale carrier air wing operations, themselves as close to actual combat conditions as reasonable prudence would permit. Over time, Strike U has conducted realistic exercises with massed Alpha Strikes by mixed forces, assembled when a full carrier air wing comes to Fallon for training. The CAG, or air wing skipper, remains in charge while Strike U provides the guidance. Routinely, an entire wing will mount a power-projection exercise, demonstrating how an Alpha Strike can assault a package of land targets and learning in the process.

The Control and Computer System which follows events as a strike is underway looks like a cross between Star Wars and a videogame parlour. Aircraft carry display and debriefing system (DDS) modules which employ telemetry to keep track of how a strike is unfolding and to be able to critique it afterwards. The learning process includes a candid rehash of 'what happened' after a strike mission is finished. Many of the exercise strike missions are flown with simulated ordnance, but often those *are* live bombs, folks: sometimes by themselves, sometimes in massed exercises with Hawkeyes helping and Tomcats guarding, Intruder crews go out over that desert and turn it to fire. The enemy's defence effort is simulated at Fallon—but the men know it won't always be. Near every ready room aboard a carrier is an Intel shop. An Intruder may be winging out over the sand and cactus at Fallon at the same time that a carrier Intel officer is sitting under a bright light scrutinizing a map of Libya's Benina airfield.

The colonel, remember?

El Dorado Canyon

The second wave of air strikes mounted against Libya on 14–15 April 1986—after conclusive proof was in hand of Khadaffi's role in the La Belle terrorist bombing—was a massive and far reaching endeavour which required much from many. Air Force tanker crews, reconnaissance and electronic warfare people, rescue forces (fortunately, not needed), the members

of REMIT flight of F-111F Aardvarks from Colonel Sam Westbrook's 48th Tactical Fighter Wing at RAF Lakenheath, England, and, of course, the same naval aircrews in Intruders and other types who had carried off Prairie Fire the month before.

Saratoga, carrying the Intruders of VA-85 and the Prowlers of VA-137, ended her line period in the Mediterranean on 5 April and headed for home. *Coral Sea* was still on the scene with the Intruders of VA-55 and the Prowlers of VAQ-135. *America* was there with Intruders of VA-34 and Prowlers of VMAQ-2.

It will be argued, eternally, whether the Air Force's F-111F portion of the operation was necessary for military reasons or whether its principal purpose was to demonstrate the backing of an ally, Great Britain,

LEFT
Secretary of the Navy John Lehman, who held the post for six years (1981–87) is also a Reservist bombardier-navigator in the Intruder and a strong proponent of the new A-6F. Lehman ordered the creation of the Naval Strike Warfare Center, known as Strike U, at NAS Fallon, Nevada in 1984 (USN)

An A-6A of Marine Corps squadron VMAT(AW)-202 pauses between operations from MCAS Cherry Point, North Carolina, on 20 July 1972 (USMC via Jim Sullivan)

which in the person of Prime Minister Thatcher courageously authorized the use of UK bases for the strike. (And it will be remembered, by all Americans in the profession of aerial arms, that an ally lacking in courage, France, declined permission to overfly its soil. It will be remembered by a lot of us for a very, very long time.) The best military argument for employing the F-111F and the one heard most often, goes like this: 'The only other airplane that can do the job is the Grumman Intruder and we just ain't got enough of them . . .'

The job: night bombing of carefully-selected targets with the most demanding requirement for precision and the most restrictive rules of engagement. Not even in a full-scale war would the need be so great for precise delivery of ordnance under the most demanding of conditions, against an enemy equipped with MiG-23s, MiG-25s, SAMs, and ZSU-23/4 anti-aircraft guns. In short: schooling at the RAGs and at Strike U lay behind and it was graduation time . . .

It is outside the scope of this work to pay tribute to the thousands of brave men who worked on, maintained, loaded and flew the KC-10s, KC-135s, rescue aircraft, electronics platforms, EF-111A and F-111F Aardvarks, Sea Kings, Hawkeyes, Whales, Corsairs and Hornets which made possible the massive El Dorado Canyon operation. But outside the scope or not, the record must be set straight on two points:

Point one: the United States set forth to attack

TOP
A-6A 152623 of VMA(AW)-121 guards its place on the
line at Cherry Point on 13 April 1971
(Jim Sullivan)

ABOVE
A-6A 152587 of VMA(AW)-121 'Moonlighters' folded up
at Cherry Point on 13 April 1971
(Jim Sullivan)

*A-6A 152620 of VMAT (AW)-202 parked during a visit
to MCAS Beaufort, South Carolina on 12 June 1971*
(Frank R MacSorley)

*KA-6D 152618 of VA-65 'Tigers' passes the time at NAS
Oceana in June 1974*
(Jerry Geer)

TOP LEFT
A-6E TRAM, bureau number 155687, modex AE-500,
carrying a hefty load of Mk 82 500-lb (227-kg) bombs
(Grumman)

LEFT
Intruding. Marine A-6Es of VMA(AW)-533 'Hawks' fly
in loose battle formation armed with retarded Snakeyes. The
nearest aircraft, bureau number 152614, modex ED-504,
lacks the TRAM turret. The two trailing aircraft are
EA-6Bs
(Grumman)

ABOVE
EA-6A 151597 of VMCJ-3 basks in the sunshine at NAS
Miramar, California, on 22 January 1974
(Robert Lawson)

OVERLEAF
KA-6D 151787, modex NH-521, of VA-52 'Knight Riders'
displays a full complement of external fuel tanks during a
visit to Richards-Gebour AFB, MO, in July 1977
(Clyde Gerdes)

TOP LEFT
*EA-6A 151596 of VMCJ-2 'Playboys' cruising through
practically clear skies in 1966*
(USMC)

LEFT
*EA-6A 151596 pictured rather later in its career being
readied for flight on the ramp at Pax River, 21 April 1971.
Note the Shrike anti-radiation missile (ARM) on the
outboard pylon*
(R Besecker/C Gerdes Collection)

ABOVE
*The clouds begin to gather over Cherry Point as A-6A
151593 waits for some kind aviator to take her up on 20 July
1972. This particular machine belongs to VMAT(AW)-202*
(Jim Sullivan)

KA-6D 151576, modex NK-516, of VA-196 'Main Battery' loiters above cloud in expectation of a thirsty naval airplane needing precious gas. The drogue housing is clearly visible under the rear fuselage
(Grumman)

legitimate targets including the command authority in Libya but *not* to assassinate Colonel Khadaffi. Notwithstanding a thorough analysis by journalist Seymour Hersh which claims the the colonel himself was the target, he wasn't. If he had been, we would have gotten him.

Point two: although some other recent US military operations in recent years have been less than successful (the Lebanon strike being one), the evidence is clear that El Dorado Canyon delivered the bombs where they were supposed to go, with a degree of accuracy and precision unparalleled in history. The Libyans showed the world television footage of alleged American bombing of civilians, and BBC journalist Kate Addie earned a place in our hearts right next to Jane Fonda by swallowing it whole. We were even accused of dropping a bomb near the French embassy in Tripoli, which we didn't even though that nation's decision makers enjoyed our affections just as much as Addie. Although there is some genuine evidence that a couple of bombs from one aircraft went astray, *virtually all damage to civilian casualties was caused by Khadaffi's own SAMs and Triple-A coming down on his own people.* Indeed, as even Addie was able to perceive and in fact acknowledged, Libyan missiles and anti-aircraft fire were whooshing around in the skies for hours before and after the brief, surgical visit by Aardvarks and Intruders.

The second attempt to send a message to Colonel Khadaffi began when the first of no fewer than thirty-two KC-10A Extender tankers began arriving at RAF Mildenhall and RAF Fairford, England, on 11 April 1986. By Monday 14 April, no fewer than thirty-four KC-135A, KC-135E and KC-135Q tankers had also deployed to these bases, while

activity at the F-111F base, Lakenheath, intensified. Air Force public relations officer Lt Col Doug Hershey and his staff told all that the feverish activity was connected with a peacetime exercise, Salty Nation. But by the evening of 14 April, it was evident that more than a training exercise was involved.

The actual strike force included six KC-135A tankers launched from Mildenhall at 5:45 pm, followed at 6:00 pm by the first of ten KC-10s. Meanwhile, fully laden with bombs, twenty-four F-111F Aardvarks launched from Mildenhall. These formed into groups of four, with REMIT Flight in the lead, and headed southwest. Five EF-111A Ravens soon followed. Six of the F-111Fs and two of the Ravens were spares, but the remainder of the force was en route to Libya by a long and circuitous means, refuelling first at Land's End, a second time off the Portuguese coast, going *around* Europe and into the Mediterranean for a third refuelling, and carrying their 500-lb (227-kg) Paveway laser-guided bombs towards Tripoli with an anticipated TOT (time over target) of midnight.

America and *Coral Sea* were assigned targets in eastern Libya, specifically the Al Jumahiriya barracks and the airport at Benghazi, while the F-111F force moved toward the capital. Beginning around 8:30 pm as dusk fell over the carrier force, *Coral Sea* sent eight Intruders and six Hornets into action. *America* despatched six Intruders and six Corsairs.

Going Downtown

With the EF-111A Ravens jamming Libyan radars from a stand-off position and the indispensable EA-6B Prowler performing the same mission closer-in, the Hornets and Corsairs approached the coast and fired HARM missiles at radar sites. At almost the same instant, roughly 11:55 pm, F-111Fs went into Tripoli while the combined force of A-6E Intruders from the two carriers crossed the beach towards Benghazi. Libyan missiles and Triple-A were sending stuff up into the air all night long, the anti-aircraft defences every bit as active both before and after the actual ingress of the strike forces. (The record book for the 'Warhorses' of VA-55 would show that their Cdr Rod Weber's Intruders flew 286 consecutive sorties during the Libyan operations, including the trip to Benghazi, without incident.)

The highly-accurate Air Force strikes also resulted in the only combat casualties of the operation, the F-111F crew consisting of Captain Fernando Ribas-Dominicci, pilot, and Captain Paul Lorence, weapon systems officer. It is thought that they were hit by Triple-A fire. Their F-111F went into the sea and, despite a massive effort, no rescue or recovery was possible.

According to one report, the HARM firings by Hornets and Corsairs were so effective that, when the Intruder force approached Benghazi, a Libyan commander was overheard to say that his SAM radars were inoperative, unable to guide missiles against the Intruders. Carrying 500-lb (227-kg) and 750-lb (340-kg) bombs, the Intruders began their low level runs at one minute past midnight, 15 April, 1986. MiG-23 fighters at Benina airfield never got off the ground and no fighters ever did challenge the Intruders. A number of aircraft were destroyed on the ground and the barracks complex was wasted by the precise Intruder strike. The destroyed aircraft are thought to have included several MiG-23s, two Mil Mi-8 *Hip* helicopters and a Fokker F.27 transport. By forty-five minutes past midnight, the Intruders had recovered aboard *America* and *Coral Sea*—with no damage and no casualties.

Was the message received? A year after El Dorado Canyon, no major terrorist incident was being attributed to the unpredictable Colonel Khadaffi. It was never possible to tell what might happen next in a complex, troubled world, but to all appearance the message had finally gotten through, loud and clear. Perhaps, all of us hoped, it would not be necessary to visit the Colonel again. The courage of British allies, the rather different viewpoint of the French, and the valour of the men who flew the missions against Libya all came together to prove the point which has been made again and again in this series about air combat: in the final years of this century, the only issue of importance is whether free men possess the willingness to employ force of arms when need is. The EA-6B Prowler, the A-6E TRAM Intruder, and all of the participants in the successful El Dorado Canyon effort demonstrated that even if the world is complex—and at times ambiguous—there still exist brave men who understand freedom, understand tyranny. None more than men in the Navy's Intruder community will remember the contribution made by the Air Force's Ribas-Dominicci and Lorence—all honour to their names.

Future Intrusions

Not merely troubled, not merely complex, the world is also fast-changing. As this volume went to the presses, it became clear that the EA-6B Prowler, described throughout this narrative as unarmed, was by 1987 being equipped with the AGM-88A HARM missile to give it a more accurate role in countering SAM missile defences.

Prowlers and Intruders will almost certainly be guarding the Fleet well into the twenty-first century. Just as they will fight battles against foreign adversaries, so, too, must they survive battles in the ever more difficult arena of the American defence budget. In fiscal year 1988, Capitol Hill had its share of legislators who wondered if the Navy really needed a future Advanced Tactical Aircraft (ATA) *and* continued production of Prowlers and Intruders. There were some who suspected that, despite the many times when the A-6 had proven itself a unique national asset, the April 1987 departure of Secretary of the Navy John Lehman was a blow to future order-book entries for the A-6. In fairness, the views of Lehman's successor, James Webb, were not known, and—a highly decorated, Annapolis-bred veteran of Marine ground combat in Vietnam—Webb was perhaps likely to be as ardent a supporter as anyone of an aircraft capable, among other duties, of precision close air support.

As this study was being completed, Grumman flew its first A-6F full-scale development (FSD) aircraft on 1 September 1987. Far more than a re-engined Intruder, although its two 10,700-lb (4853-kg) thrust General Electric F404-GE-400 powerplants would add new capability, the A-6F dated from an August 1984 contract for $1.4 million for F-14D Tomcat and A-6F Intruder development. The discussions on Capitol Hill were about the full-scale development programme slated to bring the manufacturer $379.8 million in fiscal years 1988–89 for the five initial A-6F airframes. Naval Air Systems Command wanted, and many felt the Fleet needed, 150 new airframes in the A-6F series plus a further 230 A-6Es converted to A-6F by the early 1990s.

As already noted in the discussion of Intruder variants, the A-6F will employ inverse synthetic aperture radar (ISAP) processing to enable the bombardier-navigator to classify ship targets at greater distances. Externally, the A-6F will differ in having two dorsal cooler scoops rather than one and outboard wing ordnance stations for AIM-9L Sidewinder or AIM-120A AMRAAM missiles. A

*The 'Bengals' of Marine Corps' squadron VMA(AW)-224
maintain immaculate formation during a sortie from Cherry
Point on 28 June 1967*
(USMC)

minor change is the 'deepening' of the engine cavities by 5.1 inches to accommodate the new powerplants. The A-6F will also have lightweight engine shrouds and other weight-saving characteristics. Production machines, following the first five FSD airframes, will have composite wing panels built by Boeing.

New Interior

Cockpit for the two-man crew of pilot and bombardier-navigator has been totally redesigned, with digital instruments and five readout displays for the computerized navigation and attack system. The vastly different 'ergonometric' cockpit requires the US Navy to confront cost questions arising from operating two Intruder variants at the same time in the early 1990s. The A-6E and A-6F are sufficiently

A Marine Corps' Intruder of VMA(AW)-332 'Polkadots' flares for touchdown. Landing on a long runway is a breeze compared to thumping down your airplane on a pitching, rolling, carrier deck

RIGHT
A-6E 151784 of VA-128 'Golden Intruders' makes a catapult shot from the USS Carl Vinson *(CVN-70) during a Pacific cruise in January 1986*
(USN/PHAN Charles R Solseth)

different that squadrons and crews will fly one or the other but not both. Grumman experts are consulting on the training syllabus to be offered by the Atlantic Fleet RAG (well, okay, FRS), the 'Green Pawns' of VA-42 at NAS Oceana, Virginia, which will require aircrews to stick with one Intruder or another. The A-6F first flight was scheduled a year earlier than had been originally planned.

Production A-6F Intruders with the Boeing composite wings will benefit from the ECP 1 update programme for the A-6F which will add a new night attack navigation system (NANS), a new low altitude warning system (LAWS) to provide earphone tone for ground collision avoidance, and compatibility with night-vision goggles for the crew.

Budget Battles

Each year throughout the Intruder's long career, some critic on Capitol Hill has sought to argue that it costs too much, or it's too old, or it just isn't manufactured in the right state. Each year, the Intruder has survived—and often enough, it has been called upon to prove its unique capabilities.

In fiscal 1988, financial bickering in the Congress's armed services committees was again raising questions, this time about the ATA, the Marines' AV-8B Harrier, and the A-6F Intruder. As usual, the two houses of Congress were applying different points of view: at one juncture the Senate had endorsed the Harrier, approved extra funding for ATA, and

deemed that the A-6F Intruder should not be developed. The House of Representatives, in its wisdom, cancelled AV-8B Harrier funding, reduced funds for the ATA, and retained the Navy's original programme for the next-generation Intruder. Confusing? It seems to happen every year . . .

When the two houses of Congress differ, the issue is usually patched up in conference. On the occasion now being described, it appeared likely that a conference session—resolving funding for 1988–89—would end up leaving the A-6F Intruder very much alive and kicking. Even those who love the Intruder best, and love is rarely an affection bestowed on an ugly subsonic airplane with the pointed end facing the wrong way, will admit that a newer generation of strike aircraft will one day be needed and that at some future time the ATA—in 1988 still little more than a vague concept—will supplant or replace the ungainly product of the Grumman Iron Works.

Whatever the future holds, it is unlikely that modelling enthusiasts will ever be able to put together an accurate scale replica of the A-6 Intruder in foreign insignia. Although the old tadpole shape on the outside is now very aged indeed, the internal 'fit' in the airplane is the newest and best stuff. By the very nature of its night- and bad-weather capability, the old but ever-rejuvenated Intruder remains at the fore of the state of the art. There are two few of them for Navy and Marine Corps needs, and they are too valuable, to be exported under any conditions which can now be foreseen. And the ATA? Someday, perhaps, it will fly and fight. But not some day soon.

The Challenge

And men do not fight wars with airplanes which exist as vague concepts, which indeed, in the case of ATA, do not even exist on drawing boards. Men fight with what they have.

This is not an air series, but an Air Combat Series. This is, too, the story of an airplane—its squadrons, bureau numbers, variants . . . its bumps and bulges and even its funny-looking refuelling probe and avionics birdcage. But air combat always has been about, and always will be about, men. About electricians, armourers, mechanics, maintenance men, pilots, bombardier-navigators and commanders. So again, the question, the all-important question about the Grumman Intruder and Prowler which is still something of an enigma to this author, still a rather special craft never praised enough and not always understood. In the 1990s, Americans will retain an all-volunteer force in a society where diplomat and warrior, civilian and sailor, ordinary citizen and Intruder pilot, are growing further and further apart, living in separate worlds, at times not even speaking the same language. Norman Mailer has said that our failure is that we diverge as countrymen, growing further and further apart. In a world of trouble, of complexity, of ambiguity, the question remains: when we need them, when old men like me make decisions that send young men into battle, will Intruder crews be able to do the job. Do we possess what it takes, for those occasions when only force of arms will carry the day?

Standing in the grey murk at Patuxent with Dave (Cleve) Brown, looking at Grumman's enigma with its blunt nose and supreme combat capabilities, I never had a moment's doubt. 'We have a really interesting mission', quoth Dave, 'one of the most interesting missions in all of aviation.'

No other airplane performs that mission as the Intruder does. And with people like Dave in the A-6 Intruder community, we are in good hands.

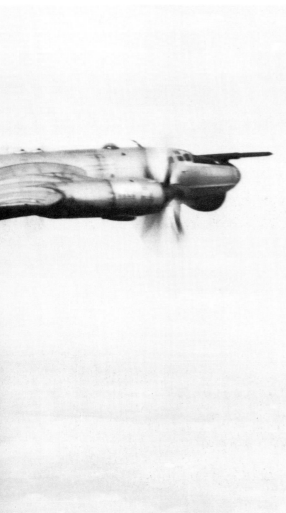

TOP LEFT
One of the newest Intruders, bureau number 262651 (side number NF-501) is flying from the decks of the oldest carrier, the USS Midway *(CV-41), scheduled to be retired from the fleet in 1992 at the age of fifty-three years! This 'low visibility' Intruder belongs to squadron VA-115*
(US Navy)

LEFT
On 11 November 1985, Marine squadron VMA(AW)-121 'chopped' to operational control of the Navy to go aboard USS Ranger *(CV-61) for a four-year period. The carrier's air wing, CVW-2, had no A-7s or F/A-18s, and the Marines sometimes found themselves flying 'fighter' missions. During* Ranger's *'surge' cruise to Korea and Japan in 1987, A-6E TRAM Intruder 151820, side number NE-414, is steered by Captain Paul Higgins towards a Russian* Bear
(USMC)

Glossary

AAA	Anti-aircraft artillery fire. Also: Triple-A
BuAer	Bureau of Aeronautics
Buno	Bureau number
CIA	Central Intelligence Agency
ECM	Electronic counter-measures
EW	Electronic warfare
G	One gravity force
NAS	Naval Air Station
SAM	Surface-to-air missile, usually referring to Soviet-made SA-2 *Guidline* missile
Tacan	Tactical air navigation system

Appendices

Appendix 1. **Intruder/Prowler bureau numbers**

	From	To	Amount	Comment		From	To	Amount	Comment
A-6A	147864	147867	4	19 converted to A-6B	A-6F	162183	162187	5	
	148615	148618	4	12 converted to EA-6A	EA-6B	(148615)			(1)
	148619	148626	(8) cx	20 converted to KA-6D		(149479)			(1)
	149475	149486	12	3 converted to EA-6B		(149481)			(1)
	149935	149958	24	12 converted to A-6C		156478	156482	5	
	151558	161594	37			158029	158040	12	
	151780	151827	48			158540	158547	8	
	152583	152646	64			158649	158651	3	
	152891	152954	64			158799	158817	19	
	152955	152964	(10) cx			159582	159587	6	
	154046	154099	(54) cx			159907	159912	6	
	154124	154171	48			160432	160432	6	
	155137	155190	(54) cx			160609		1	
	155581	155721	141			160704	160709	6	
	156994	157029	83			160786	160791	6	
		SUBTOTAL	482			161115	161120	6	
EA-6A	156979	156993	15	(Plus 12 converted)		161242	161247	6	
SUBTOTAL						161347	161352	6	
A-6E	158041	158052	12			161774	161779	6	
	158528	158539	12			161880	161885	6	
	158787	158798	12			162934	162941	8	
	159174	159185	12			163030	163035	6	
	159309	159317	9			163044	163049	6	
	159567	159581	15			163395	163406	12	
	160421	160431	11						
	160993	160998	6						
	161082	161093	12						
	161112	161114	(3) cx						
	161230	161235	6						
	161236	161241	(6) cx						
	161659	161690	32						
	161691	161694	(4) cx						
	161886	161897	(12) cx						
	162179	162182	4						
	162188	162222	35						
	163520	163530	(11) cx						

Appendix 2. **US Navy A-6 Squadrons**

Unit	Nickname	Traditional Home Port
VA-34	Blue Blasters	NAS Oceana, Virginia
VA-35	Black Panthers	Oceana
VA-42	Green Pawns	Oceana
VA-52	Knight Riders	NAS Whidbey Island
VA-55	Warhorses	Oceana
VA-65	Tigers	Oceana
VA-75	Sunday Punchers	Oceana
VA-85	Black Falcons	Oceana
VA-95	Green Lizards	Whidbey
VA-115	Arabs	Whidbey
VA-123	Professionals	Whidbey
VA-128	Golden Intruders	Whidbey
VA-145	Swordsmen	Whidbey
VA-165	Boomers	Whidbey
VA-176	Thunderbolts	Oceana
VA-196	Main Battery	Whidbey
VX-5	Vampires	NAS China Lake, Calif

Appendix 3. **US Marine Corps A-6 Squadrons**

Unit	Nickname	Home Base
VMA(AW)-121	Moonlighters	MCAS Cherry Point, NC
VMAT(AW)-202	*	Cherry Point
VMA(AW)-224	Bengals	Cherry Point
VMA(AW)-225	Vagabonds	MCAS El Toro, Calif
VMA(AW)-242	Bats	El Toro
VMA(AW)-332	Polkadots	Cherry Point
VMA(AW)-533	Hawks	MCAS Iwakuni, Japan

*VMAT(AW)-202, formed on 15 Jan 1968, did not have an assigned nickname as of 20 Dec 1979. The squadron was deactivated 30 Sept 1986

Appendix 4. **US Navy and Marine Corps EA-6B Squadrons**

Unit	Nickname	Home Base
VAQ-129	Vikings	NAS Whidbey Island, Wash
VAQ-130	Zappers	Whidbey
VAQ-131	Lancers	Whidbey
VAQ-132	Scorpions	Whidbey
VAQ-133	Wizards	Whidbey
VAQ-134	Garudas	Whidbey
VAQ-135	Black Ravens	Whidbey
VAQ-136	Gauntlets	Whidbey
VAQ-137	Rooks	Whidbey
VAQ-138	Yellow Jackets	Whidbey
VAQ-139	Cougars	Whidbey
VAQ-140		Whidbey
VMAQ-2	Playboys	MCAS Cherry Point, NC

Appendix 5. **US Navy and Marine Corps EA-6A Squadrons**

Unit	Nickname	Home Base
VMCJ-1		MCAS Iwakuni, Japan
VMCJ-2	Playboys	MCAS Cherry Point, NC
VMCJ-3		MCAS El Toro, Calif
VMAQ-2	Playboys	Cherry Point
VMAQ-4	Seahawks	NAS Whidbey Island, Wash
VAQ-33	Firebirds	NAS Norfolk, Va
VAQ-209	Star Warriors	Norfolk
VAQ-309	Axeman	Whidbey

Specifications

Grumman A-6A Intruder

Type: two-seat carrier-based all-weather strike/attack aircraft

Powerplant: two 9,300-lb (4218-kg) non-afterburning Pratt & Whitney J52-8A two-shaft turbojets

Performance: maximum speed (clean) 685 mph (1102 km/h) at sea level or 625 mph (1006 km/h) at height; service ceiling 41,660 ft (12,700 m); ferry range with external fuel (all) about 3,100 miles (4890 km)

Weights: empty 25,684 lb (11,650 kg); maximum takeoff 60,400 lb (27,397 kg)

Dimensions: span 53 ft (16.15 m); length 54 ft 7 in (16.64 m); height 15 ft 7 in (4.75 m); wing area 528.9 sq ft (49.1 m²)

Armament: five stores locations each rated at 3,600 lb (1633 kg) with maximum total load of 15,000 lb (6804 kg)

Grumman A-6E Intruder

Type: as A-6A

Powerplant: as A-6A

Performance: maximum speed (clean) 644 mph (1037 km/h) at sea level; initial climb rate (clean) 8,600 ft (2621 m)/m; service ceiling 44,600 ft (13,595 m); range with full combat load 1,077 miles (1733 km); typical ferry range 2,735 miles (4401 km)

Weights: empty 25,630 lb (11,625 kg); maximum takeoff as A-6A

Dimensions: as A-6A except height 16 ft 3 in (4.95 m)

Armament: five stores pylons each rated at 3,600 lb (1633 kg) with maximum load of 18,000 lb (8165 kg) including combinations of up to thirty 500 lb (225 kg) Rockeye Mk 20 cluster bombs and high-drag (with retarding system) or 'slick' Snakeye Mk 82 bombs; three B28 free-fall nuclear bombs; LAU-10 or LAU-32 rocket pods with three and nineteen 5 in (127 mm) Zuni rockets respectively; AGM-45 or AGM-88 HARM anti-radar missiles; six AGM-84 Harpoon anti-ship missiles; and various mines and torpedoes

Grumman EA-6B Prowler

Type: four-seat carrier-based electronic warfare platform

Powerplant: two 11,200-lb (5080-kg) thrust non-afterburning Pratt & Whitney J52-408 turbojets

Performance: maximum speed (clean) 599 mph (964 km/h) at sea level; service ceiling 39,000 ft (11,582 m)

Weights: empty 34,581 lb (15,686 kg); maximum loaded 58,500 lb (26,535 kg)

Dimensions: as A-6 except length 59 ft 5 in (18.11 m)

Armament: HARM

Index